USMLE ROAD MAP

GROSS ANATOMY

Second Edition

JAMES S. WHITE, PhD

Assistant Professor
of Cell Biology
School of Osteopathic Medicine
University of Medicine and Dentistry of New Jersey
Stratford, New Jersey

Adjunct Assistant Professor
of Cell and Developmental Biology
University of Pennsylvania School of Medicine
Philadelphia, Pennsylvania

Lange Medical Books/McGraw-Hill
Medical Publishing Division

New York   Chicago   San Francisco · Lisbon   London   Madrid   Mexico City
Milan   New Delhi   San Juan   Seoul   Singapore   Sydney   Toronto

USMLE Road Map: Gross Anatomy, Second Edition

Copyright © 2006 by The McGraw-Hill Companies, Inc. All rights reserved. Printed in the United States of America. Except as permitted under the United States Copyright Act of 1976, no part of this publication may be reproduced or distributed in any form or by any means, or stored in a data base or retrieval system, without prior written permission of the publisher.

34567890 DOC/DOC 0987

ISBN: 0-07-144516-1

ISSN: 1543-5806

This book was set in Adobe Garamond by Pine Tree Composition, Inc.
The editors were Jason Malley, Robert Pancotti, and Mary E. Bele.
The production supervisor was Sherri Souffrance.
Project management was provided by Pine Tree Composition, Inc.
The index was prepared by Sherri Detrich.
RR Donnelley was printer and binder.

This book is printed on acid-free paper

INTERNATIONAL EDITION ISBN 0-07-110313-9 Copyright © 2006. Exclusive right by The McGraw-Hill Companies, Inc. for manufacture and export. This book cannot be re-exported from the country to which it is consigned by McGraw-Hill. The International Edition is not available in North America.

CONTENTS

USING THE
USMLE ROAD MAP SERIES
FOR SUCCESSFUL REVIEW

What Is the Road Map Series?

Short of having your own personal tutor, the USMLE Road Map Series is the best source for efficient review of major concepts and information in the medical sciences.

Why Do You Need A Road Map?

It allows you to navigate quickly and easily through your anatomy course notes and textbook and prepares you for USMLE and course examinations.

How Does the Road Map Series Work?

Outline Form: Connects the facts in a conceptual framework so that you understand the ideas and retain the information.

Color and Boldface: Highlights words and phrases that trigger quick retrieval of concepts and facts.

Clear Explanations: Are fine-tuned by years of student interaction. The material is written by authors selected for their excellence in teaching and their experience in preparing students for board examinations.

Illustrations: Provide the vivid impressions that facilitate comprehension and recall.

 Clinical Correlations: Link all topics to their clinical applications, promoting fuller understanding and memory retention.

 Clinical Problems: Give you valuable practice for the clinical vignette-based USMLE questions.

 Explanations of Answers: Are learning tools that allow you to pinpoint your strengths and weaknesses.

To my wife, Kim, and my daughters, Kate and Kristen,
and to the memory of Sasha Malamed
and Irwin Baird; both extraordinary teachers, mentors, and colleagues.

Acknowledgments

I am indebted to Harriet Lebowitz, Janet Foltin, Regina Y. Brown, and Mary McCoy for their efforts in preparing the manuscript of the First Edition. Thanks also go to my colleagues David Seiden, PhD, and George Mulheron, PhD, for their critical input on the First Edition, and to Douglas Jaffe, RN, medical student at the School of Osteopathic Medicine at the University of Medicine and Dentistry of New Jersey, Stratford, New Jersey, for his editorial efforts in preparation of the Second Edition.

CHAPTER 1
FUNDAMENTALS

I. The **skeletal system** consists of bones of the axial and appendicular skeleton.

   A. The **80 bones of the axial skeleton** include the bones of the skull, vertebral column, ribs, sternum, and the hyoid.

   B. The **126 bones of the appendicular skeleton** include bones of the upper and lower limbs and the pectoral and pelvic girdles, which link the limbs to the axial skeleton.

   C. **Bones** are composed of an outer region of compact or dense bone and an inner region of spongy bone containing spaces and a cavity of marrow. Bones are covered on their external surfaces by periosteum, which is continuous with deep fascia.

FRACTURES

CLINICAL CORRELATION

- Bones may **fracture** as a result of trauma or may undergo atrophy from either disuse or **osteoporosis**, a loss of bone mass.

- Fractures result **in avascular necrosis**, a loss of bone tissue caused by disruption of arterial blood supply.

II. **Joints** are formed where 2 bones are joined together by nonosseous elements.

   A. In **synarthrodial, or solid, joints**, collagenous or fibrocartilaginous connective tissue occupies the space between the bony elements; little movement is permitted.

   B. In **diarthrodial, or synovial, joints**, a space with synovial fluid separates the articulating surfaces of the opposed bones; freedom of movement is permitted.

III. The **muscular system** consists of skeletal muscle, cardiac muscle, and smooth muscle.

   A. Skeletal Muscle

   1. **Skeletal muscle** consists of individual elongated, unbranched, striated muscle cells or fibers, which, when depolarized, can contract to about half their length.

   2. Skeletal muscle attaches to bone by way of a tendon, an aponeurosis, or a fleshy attachment.

   3. Skeletal muscle is under **voluntary or reflexive control** by way of motor units composed of a skeletal motor neuron and all of the muscle cells it innervates. Depolarization of a skeletal motor neuron results in a contraction of all of the muscle cells in the motor unit.

a. **Large muscles**, such as the quadriceps femoris and gluteus maximus that produce gross movements consist of large motor units.

b. **Small muscles**, such as intrinsic hand muscles, that produce fine movements consist of small motor units.

4. Skeletal muscle is **innervated at the neuromuscular junction**, the site at which the axon of a motor neuron forms a synapse with an end plate of a muscle fiber (Figure 1–1).

5. **Acetylcholine** released by the axon terminal binds to cholinergic receptors (ion channels) in the membrane of the end plate. Opening of these channels permits an influx of sodium ions, which depolarizes the muscle membrane.

MYASTHENIA GRAVIS (FIGURE 1–1)

In **myasthenia gravis**, antibodies attack the acetylcholine receptors, resulting in defective neuromuscular transmission.

• **Muscles innervated by cranial nerves** are primarily affected, with ocular muscles almost always included. Weakness increases with use.

• Patients have bilateral ptosis and horizontal diplopia and may have dysphagia, dysarthria, and weakness in chewing and in the muscles of facial expression. Proximal limb muscles may be affected. Cardiac and smooth muscle are spared.

• Many patients with myasthenia gravis **have thymic hyperplasia** or a **thymoma.**

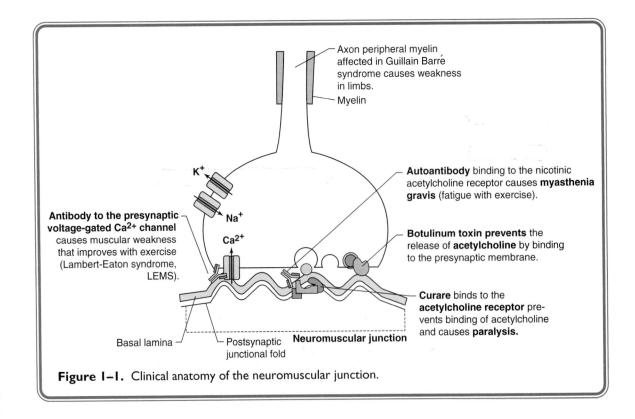

Figure 1–1. Clinical anatomy of the neuromuscular junction.

LAMBERT-EATON SYNDROME (FIGURE 1–1)

CLINICAL
CORRELATION

In **Lambert-Eaton syndrome,** there is **an immunologic disorder of calcium ion channels** in the nerves at the end plate.
- Proximal muscles in limbs are primarily affected; muscles innervated by cranial nerves are spared. Repetitive contractions of affected muscles temporarily increase strength.
- Lambert-Eaton syndrome is associated with **small cell carcinoma of the lung** in most patients with this syndrome.

   B. Cardiac Muscle
      1. Cardiac muscle forms the myocardium of the heart and consists of **highly branched, striated fibers** interconnected by intercalated disks.
      2. This muscle is under **involuntary control;** contraction is initiated by the cardiac conduction system and modified by autonomic nerves.
   C. Smooth Muscle
      1. Smooth muscle consists of single, nonstriated fibers and is found in the walls of visceral structures and blood vessels, in the orbit, and in skin.
      2. Smooth muscle produces **rhythmic contractions** to move substances (peristalsis), restricts flow in vessels, and forms sphincters.
      3. Smooth muscle is under **involuntary control** by autonomic nerves.

IV. The **nervous system** consists of the **central nervous system** (CNS) and the **peripheral nervous system** (PNS). Both parts of the nervous system consist of neurons and glial or supporting cells.
   A. CNS and PNS
      1. The **CNS** includes the brain in the cranial cavity and the spinal cord in the vertebral canal.
      2. The **PNS** is composed of spinal nerves, cranial nerves, autonomic nerves, and their branches.
   B. Neurons
      1. Neurons typically consist of a nerve cell body and 2 types of cytoplasmic processes—a single axon and multiple, branched dendrites.
      2. Neurons have **cell bodies** that are found in either a ganglion or a nucleus. A **ganglion** is a collection of neuron cell bodies situated in the PNS. A **nucleus** is a collection of neuron cell bodies inside the brain or spinal cord (Figure 1–1).
      3. There are only **2 types of ganglia:** sensory and autonomic.
         a. Sensory ganglia contain cell bodies of either pseudounipolar or bipolar sensory neurons. Both types of neurons have a central process that acts like an axon in conducting impulses into the CNS and a peripheral process that acts like a dendrite in conducting impulses from a sensory receptor toward the cell body. There are no synapses in sensory ganglia.
         b. Autonomic ganglia, which contain motor cell bodies, are synaptic sites where an impulse is transmitted from the axon of a preganglionic autonomic neuron to the dendrites or cell body of a postganglionic neuron.
         c. Neurons in ganglia are derived from **neural crest ectoderm.**
      4. Nuclei contain a wide variety of functionally or anatomically similar neurons.
         a. Skeletal motor neurons and preganglionic autonomic neurons, which give rise to axons in nerves, are found in nuclei.
         b. Neurons in nuclei are derived from the neural tube.

5. **Axons** and central and peripheral sensory processes in the CNS and PNS may be **myelinated.** Myelinated axons with thicker myelin sheaths conduct impulses faster than thinly myelinated or unmyelinated axons.

  a. **Oligodendrocytes** are glial cells that form myelin for parts of multiple axons in the CNS. Although oligodendrocyte actually means *having few processes,* they myelinate multiple axons.

  b. **Schwann cells** are glial cells that form myelin for axons or processes in the PNS. Each Schwann cell forms myelin for only a single segment of a single axon.

6. **CNS axons** with myelin sheaths formed by oligodendrocytes **do not regenerate if cut.** Axonal regeneration may be inhibited by proliferation of glial cells at the site of the lesion, inflammation, or lack of factors secreted by oligodendrocytes.

7. **Myelinated axons in the PNS** have the capacity to **regenerate** down an endoneurial sheath formed by Schwann cells. Peripheral regeneration occurs at a rate of about 1–4 mm/day, but the axons may grow back to the wrong target or may not regenerate completely.

MULTIPLE SCLEROSIS

In **multiple sclerosis (MS)**, **both sensory and motor systems** containing axons with myelin formed by oligodendrocytes undergo an **inflammatory reaction** that impairs or blocks impulse transmission.

- **Multiple lesions** appear over time, but the signs and symptoms may undergo exacerbation and remission.
- **Sensory and motor deficits** may be seen in all areas of the body. Usually 2 or more CNS neural systems are affected in separate attacks.
- The only nerve affected is the **optic nerve** (CN II) because all of the myelin sheaths of its axons are formed by oligodendrocytes. Optic neuritis is the presenting sign of MS in 40% of diagnosed patients.
- Other spinal and cranial nerves are not affected because their myelin is formed almost entirely by Schwann cells.

GUILLAIN-BARRÉ SYNDROME

In **Guillain-Barré syndrome**, **myelin** formed by Schwann cells in the PNS undergoes an **acute inflammatory reaction** after a respiratory or gastrointestinal illness. This reaction also impairs or blocks impulse transmission.

- **Motor axons** are always affected, producing weakness in the limbs. Weakness of cranial nerve innervated muscles (most commonly those innervated by CNs VI and VII) or respiratory muscles may be seen. Sensory deficits are mild or absent.

SCHWANNOMAS

Benign encapsulated **schwannomas of the vestibulocochlear nerve** (CN VIII) may develop in Schwann cells.

- These tumors compress CN VIII and affect hearing and balance. **Large acoustic schwannomas** may also compress the facial nerve (CN VII) and the trigeminal nerve (CN V).
- **Bilateral acoustic schwannomas** occur in patients with neurofibromatosis Type 2.

  C. **Spinal Nerves (Figures 1–2, 1–3)**

  1. There are 31 pairs of **spinal nerves:** 8 cervical, 12 thoracic, 5 lumbar, 5 sacral, and 1 coccygeal. All branches of spinal nerves contain both motor and sensory fibers.

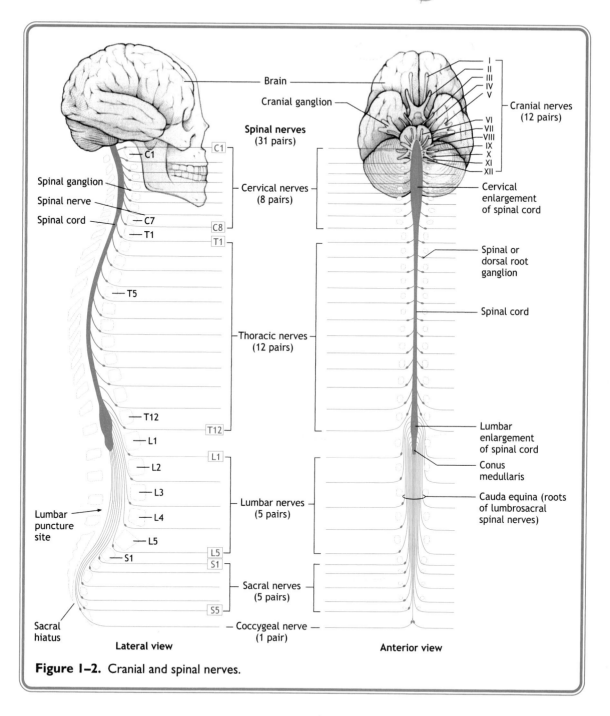

Figure 1–2. Cranial and spinal nerves.

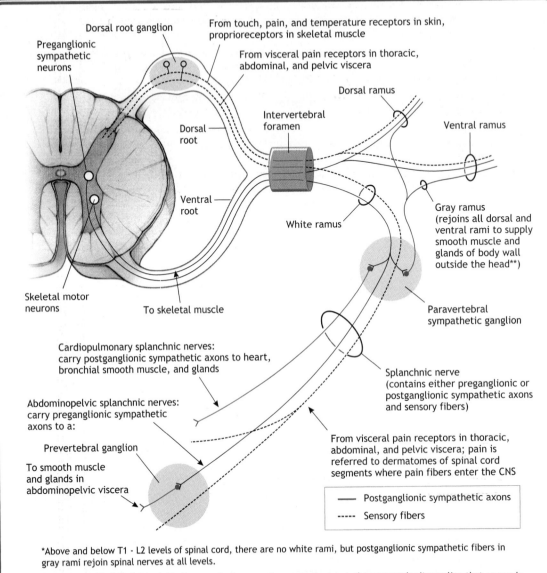

Figure 1–3. Components of a spinal nerve from T1–L2.

Within the figure:

Dorsal root ganglion

Preganglionic sympathetic neurons

From touch, pain, and temperature receptors in skin, proprioreceptors in skeletal muscle

From visceral pain receptors in thoracic, abdominal, and pelvic viscera

Dorsal ramus

Intervertebral foramen

Ventral ramus

Dorsal root

Ventral root

White ramus

Gray ramus (rejoins all dorsal and ventral rami to supply smooth muscle and glands of body wall outside the head**)

Skeletal motor neurons

To skeletal muscle

Paravertebral sympathetic ganglion

Cardiopulmonary splanchnic nerves: carry postganglionic sympathetic axons to heart, bronchial smooth muscle, and glands

Splanchnic nerve (contains either preganglionic or postganglionic sympathetic axons and sensory fibers)

Abdominopelvic splanchnic nerves: carry preganglionic sympathetic axons to a:

Prevertebral ganglion

To smooth muscle and glands in abdominopelvic viscera

From visceral pain receptors in thoracic, abdominal, and pelvic viscera; pain is referred to dermatomes of spinal cord segments where pain fibers enter the CNS

——— Postganglionic sympathetic axons

----- Sensory fibers

*Above and below T1 - L2 levels of spinal cord, there are no white rami, but postganglionic sympathetic fibers in gray rami rejoin spinal nerves at all levels.

**Head is supplied by postganglionic sympathetic axons from superior cervical (paravertebral) ganglion that course in periarterial plexuses on branches of internal and external carotid arteries.

Splanchnic nerves emerge from paravertebral ganglia at all levels to carry autonomic and sensory fibers to or from visceral structures.

2. Spinal nerves are **formed by the union of a dorsal root and a ventral root** that emerge segmentally from the spinal cord and join approximately at the level of an intervertebral foramen.

3. Nerve fibers in the **dorsal root** consist of afferent, or sensory, processes. Each dorsal root has a dorsal root ganglion that contains sensory cell bodies of all the functional modalities associated with that individual spinal nerve.

4. Nerve fibers in the **ventral root** contain efferent, or motor, axons. The cell bodies of motor axons are found in gray matter inside the spinal cord.

5. Each spinal nerve divides into 2 branches: **a dorsal ramus and** a **ventral ramus**, each of which carry both motor and sensory fibers (Figure 1–3).

6. Dorsal rami
   a. Dorsal rami **innervate the skin of the medial two thirds of the back** from the vertex of the skull, corresponding to a coronal plane through the external auditory meatuses, to the coccyx.
   b. They also **innervate the deep (intrinsic) muscles of the back** that act on the vertebral column.
   c. Dorsal rami **innervate the zygapophyseal (facet) joints** between articular processes of vertebra.
   d. Few branches of dorsal rami form nerves that have names. The few dorsal rami that are named are the suboccipital nerve (dorsal ramus of C1), the greater occipital nerve (dorsal ramus of C2), and cluneal nerves that provide cutaneous innervation to the gluteal region.

7. Ventral rami
   a. Ventral rami contain more motor and sensory fibers than dorsal rami and have a more widespread distribution.
   b. Ventral rami **innervate the skin, muscles, and joints** in the ventrolateral aspects of the neck and trunk and in both extremities.
   c. They **intermingle to form plexuses of fibers** (except those from the T2-11 spinal nerves) that innervate the extremities and the neck.
   d. They **communicate with the sympathetic chain** of paravertebral (autonomic) ganglia by white and gray communicating rami.
   e. Virtually all ventral rami **form nerves that have names.**

8. Cutaneous branches of dorsal and ventral rami, except for the C1 spinal nerve, supply a specific **dermatome**, the area of skin supplied by the branches of a single spinal nerve. The dermatomes supplied by adjacent spinal nerves may overlap; consequently, a lesion of a single spinal nerve may not result in a cutaneous sensory loss (Figure 1–4).

9. Muscular branches of dorsal and ventral rami supply a specific **myotome**, the muscle mass supplied by the branches of a single spinal nerve.

10. Splanchnic branches of ventral rami from the T1-L2 and S2-4 spinal nerves **carry autonomic and sensory fibers** to organs in the thorax, abdomen, and pelvis.

D. Cranial Nerves (Figure 1–2)
   1. Cranial nerves consist of **12 pairs of nerves** that emerge from the brain (usually from the brainstem) in the cranial cavity and exit the cranial cavity through a foramen, or opening, in the skull.
   2. They mainly **supply structures found in the head and** in the **visceral part of the neck;** the vagus nerve (CN X) also supplies the heart, lungs, and gastrointestinal structures in the thorax and abdomen.

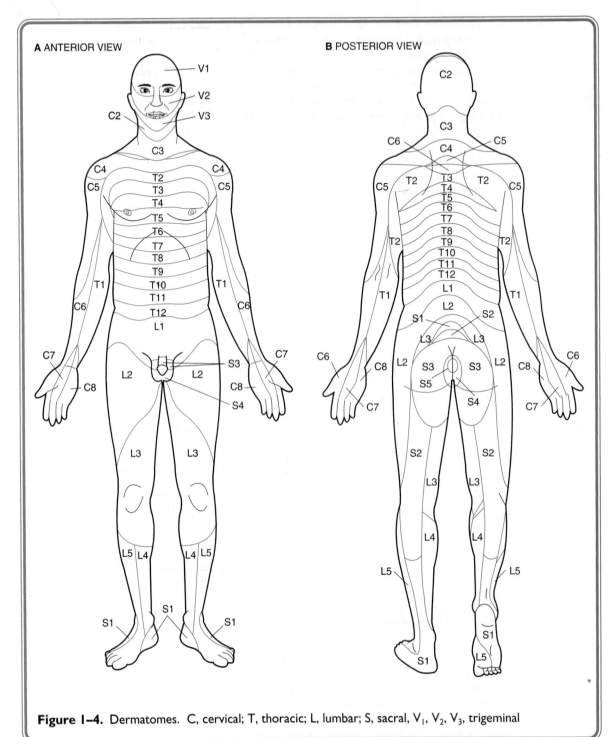

Figure 1–4. Dermatomes. C, cervical; T, thoracic; L, lumbar; S, sacral, V_1, V_2, V_3, trigeminal

3. Cranial nerves have Roman numeral designations (I–XII) and special names generally indicating the regions or structures supplied (eg, CN I: olfactory, or smell; CN IX: glossopharyngeal, or tongue and pharynx) or are named for some descriptive feature about the nerve (eg, CN X: vagus, or wanderer; CN V: trigeminal, or 3 twins).

4. Cranial nerves **have different combinations of motor or sensory fibers;** at their point of entry into or exit from the brain, 3 cranial nerves (I, II, and VIII) can be classified as sensory, 5 as motor (III, IV, VI, XI, and XII), and 4 as mixed (V, VII, IX, and X).

5. The 4 cranial nerves that are **mixed** (V, VII, IX, and X) **innervate structures derived from pharyngeal arches.**

E. **Autonomic Nerves (Figure 1–5 and Table 1–1))**

1. Autonomic nerves branch from cranial or spinal nerves and provide **motor innervation of smooth muscle, glands, and cardiac muscle.** They carry sensory information from visceral structures.

2. Autonomic nerves are organized **into sympathetic and parasympathetic divisions,** which use 2 neurons in series to provide motor innervation to target tissues.

   a. The cell body of the first neuron (the preganglionic neuron) is found in the CNS, and its axon terminates by synapsing with a second neuron (the postganglionic neuron) in an autonomic ganglion.

   b. The postganglionic axon leaves the ganglion to innervate smooth muscle, a gland, or the heart.

3. The **sympathetic division** functions as an emergency or catabolic system involved in fight-or-flight responses; sympathetic axons are widely distributed throughout the body (Table 1–1).

   a. **Preganglionic sympathetic neuron** cell bodies are found in the thoracic and upper lumbar segments of the spinal cord from T1 through L2 levels.

   b. All preganglionic sympathetic axons leave the cord and course in the ventral roots and ventral rami of the T1 through L2 spinal nerves. They then exit **in white rami communicantes** to synapse in a **paravertebral ganglion** in the sympathetic trunk or pass through the trunk to synapse in a **prevertebral sympathetic ganglion.**

   c. **Postganglionic sympathetic axons** from paravertebral ganglia course in **gray rami,** which rejoin all spinal nerve branches to innervate smooth muscle in the walls of blood vessels and sweat glands in the body wall, in both extremities, and in the neck.

   d. Postganglionic sympathetic axons from paravertebral ganglia also course **in cervical and upper thoracic (or cardiopulmonary) splanchnic nerves** to innervate smooth muscle and glands in the lungs, cardiac muscle, and the conduction system of the heart.

   e. **Prevertebral (or collateral) ganglia** are found below the diaphragm. They are situated mainly around the origins of branches of the abdominal aorta anterior to the bodies of the vertebrae and include the celiac, superior mesenteric, aorticorenal, and inferior mesenteric ganglia.

   f. Preganglionic sympathetic axons reach prevertebral ganglia in **lower thoracic, lumbar, or abdominopelvic splanchnic nerves.** Lower thoracic splanchnic nerves also synapse with chromaffin cells in the adrenal medulla, which is essentially a prevertebral ganglion.

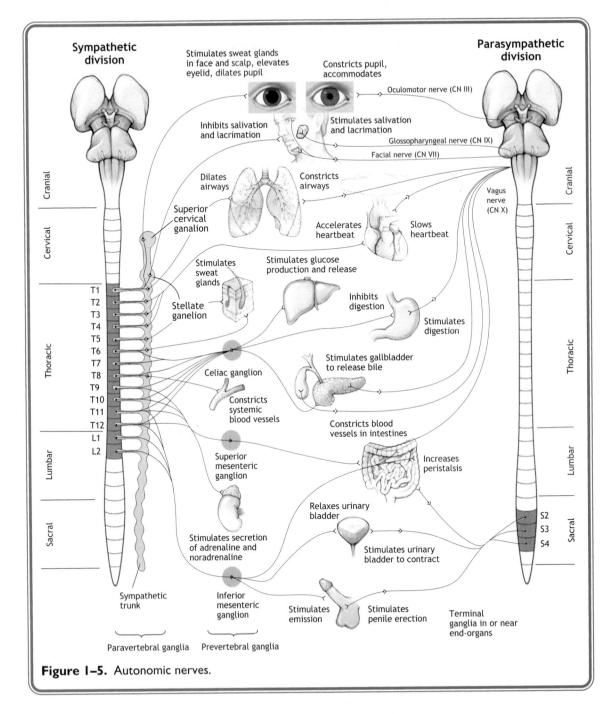

Figure I–5. Autonomic nerves.

Table 1–1. Comparison of sympathetic and parasympathetic components.

Features	Sympathetic	Parasympathetic
General	Energy mobilization; catabolic, mobilizes entire body in fight-or-flight response	Conservation; rests and digests
Anatomic		
Location of preganglionic cell bodies (in CNS)	In intermediolateral cell column of thoracolumbar spinal cord from T1-L2	In nuclei of cranial nerves in brainstem or in sacral spinal cord from S2-4
Axons exit from CNS in	Ventral roots, then ventral rami of spinal nerves from T1-L2; leave in white rami to enter paravertebral trunk	In CNs III, VII, IX, or X or in ventral roots and rami of spinal nerves S2, S3, S4
Names of ganglia	Paravertebral (in sympathetic trunk), or Prevertebral (found near origins of abdominopelvic arteries)	Terminal Ciliary, pterygopalatine, submandibular, otic
Distribution of postganglionic axons	They go everywhere; leave paravertebral ganglia to rejoin all spinal nerves in gray rami; leave prevertebral ganglia to course on abdominopelvic arteries or in visceral plexuses	Restricted to head, neck, and viscera; leave named ganglia to course with branches of CN V; leave terminal ganglia to course in visceral plexuses
Splanchnic nerves (carry autonomic axons and visceral sensory fibers)	Cervical and upper thoracic splanchnic nerves: contain postganglionic sympathetic axons from paravertebral ganglia (middle cervical to T5) Lower thoracic (greater, lesser, lowest) and lumbar splanchnic nerves: contain preganglionic sympathetic axons from T5 to L2 spinal cord segments; synapse in prevertebral ganglia	Pelvic splanchnic nerves; carry preganglionics from S2, S3, S4; synapse in abdominopelvic terminal ganglia
Chemical		
(all preganglionic neurons use acetylcholine as their neurotransmitter)	Postganglionic neurons use norepinephrine (except for axons to eccrine sweat glands, which use acetylcholine)	Postganglionic neurons use acetylcholine

CNS, central nervous system; CN, cranial nerve.

    g. Postganglionic axons from **prevertebral ganglia** supply glands and smooth muscle in organs in the abdomen and pelvis.

    h. All preganglionic sympathetic axons use **acetylcholine** as a neurotransmitter. All postganglionic sympathetic axons use **norepinephrine** as a neurotransmitter, except for those innervating eccrine sweat glands and erectile tissues, which use acetylcholine.

4. The **parasympathetic division** functions to conserve energy and restore body resources. The distribution of parasympathetic axons is restricted; they innervate 2 smooth muscles in the orbit, mucous, and salivary glands in the head, and smooth muscle and glands in the walls of thoracic and abdominopelvic viscera.

    a. The **cell bodies of preganglionic parasympathetic neurons** are situated either in the brainstem and exit the brain in 1 of 4 cranial nerves or are found at sacral spinal cord segments S2, S3, and S4 and exit the spinal cord with the ventral roots of the S2 through S4 spinal nerves (Figure 1–5).

    b. Four cranial nerves (III, VII, IX, and X) carry preganglionic parasympathetic axons out of the brainstem. These axons synapse with specific, named parasympathetic ganglia in the head. Preganglionic parasympathetic axons in the vagus (CN X) nerve synapse **in terminal ganglia** near the heart and lungs in the thorax and in the walls of the gastrointestinal tract in the foregut and midgut (from the pharynx to approximately two-thirds of the way across the transverse colon).

    c. **Sacral preganglionic parasympathetic axons** exit the spinal cord in sacral spinal nerves S2 through S4 but leave the ventral rami of these nerves in **pelvic splanchnic nerves.** Pelvic splanchnic nerves synapse with terminal ganglia in the hindgut (the distal third of the transverse colon, descending colon, sigmoid colon, and rectum), in the bladder, and in the pelvis.

    d. All preganglionic and postganglionic parasympathetic axons use **acetylcholine** as a neurotransmitter.

5. **Visceral afferent fibers**, which convey physiologic sensations or visceral pain, course with sympathetic and parasympathetic nerves (Figure 1–3).

6. **Visceral afferents that convey physiologic sensations other than pain** course with **parasympathetic axons** in the vagus or pelvic splanchnic nerves.

    a. These sensations arise from mechanoreceptors that respond to distension in the walls of respiratory structures, gastrointestinal tract, and bladder, and baroreceptors in arteries that respond to changes in blood pressure.

    b. They also arise from chemoreceptors that respond to changes in the concentration of hydrogen ions in the stomach, in the partial pressure of oxygen and carbon dioxide in the blood, and to changes in the pH of the blood.

7. **Visceral pain fibers** usually **accompany sympathetic nerves** back to the CNS and enter at thoracolumbar spinal cord levels T1 through L2 and refer pain to these dermatomes (Figure 1–3).

    a. These sensations arise from visceral pain or nociceptors stimulated by excessive distension of a part of the gastrointestinal tract or the bladder.

    b. Pain can also be caused by strong contractions of smooth muscle in the wall of a visceral structure such as the uterus.

    c. These sensations may also be caused by a lack of oxygen to cardiac muscle.

8. Referred pain
    a. **Referred pain** corresponds to **sensations that originate from thoracic, ab-dominal, or pelvic organs** but is perceived consciously in dermatomes on the surface of the body.
    b. Referred pain is **dull and poorly localized.**
    c. It is commonly **felt or referred to the dermatome** corresponding to the same spinal cord level where the pain afferents innervating that organ enter the CNS.
    d. Because most visceral pain fibers course with sympathetic nerves and enter the CNS at the level of T1 through L2, **most visceral pain is referred to the T1 through L2 dermatomes.**

HORNER'S SYNDROME

- *Horner's syndrome* may be caused by a *lesion* in either *preganglionic or postganglionic sympathetic neurons* that innervate sweat glands and blood vessels in the face and scalp and 2 smooth muscles in the orbit. The smooth muscles elevate the upper eyelid and dilate the pupil.
- Patients have *anhydrosis* (inability to sweat on the corresponding side of the face), *ptosis* (drooping of the upper eyelid), and *miosis* (constriction of the pupil).

SHY-DRAGER SYNDROME

- *Shy-Drager syndrome* results from *degeneration of preganglionic sympathetic and parasympathetic neurons* in the brainstem and spinal cord and degeneration of neurons in most ganglia. The syndrome may be combined with loss of other non-autonomic CNS neurons.
- Patients may experience *impotence*, *urine retention* with increased urinary frequency at night, *dizziness on standing*, *blurred vision*, and *inability to sweat*.

HIRSCHSPRUNG'S DISEASE (AGANGLIONIC MEGACOLON)

- *Hirschsprung's disease* is caused by a *failure of neural crest cells either to migrate* into the wall of the descending colon, sigmoid colon, or rectum or to differentiate into terminal parasympathetic ganglia in these areas.
- There is an *absence of peristalsis* in the affected segment and a distended bowel proximal to that segment.

F. Reflexes
    1. **A reflex is an automatic motor response to a sensory stimulus** and includes, at a minimum, 1 sensory neuron and 1 motor neuron that communicate at a synapse.
    2. In response to a stimulus, sensory fibers enter the CNS and stimulate either a skeletal motor neuron or an autonomic motor neuron.
        a. In **muscle stretch reflexes**, muscle spindles in skeletal muscles are stimulated by stretch, and skeletal motor neurons facilitate a reflex contraction of that same muscle.
        b. In **autonomic reflexes**, sensory stimuli and autonomic neurons facilitate a reflex contraction of smooth muscle, the secretion of a gland, or a change in the rate and force of contraction of cardiac muscle.
        c. In **cranial nerve reflexes**, sensory and either skeletal or autonomic neurons in 1 or more cranial nerves are utilized in the pupillary light reflex, the blink reflex, and the gag reflex.

NERVE LESIONS

In **irritative** lesions, nerve fibers are compressed and the firing of the irritated sensory or motor neurons is altered.

- Irritative lesions of sensory fibers commonly result in reduced sensation (**hypesthesia**) or altered sensation (**paresthesia**).
- Irritative lesions of motor fibers may result in weakness (**paresis**) of skeletal muscles. In **destructive** lesions, nerve fibers are severely compressed or severed, resulting in a loss of the ability of nerves to conduct impulses.
- **Destructive lesions of sensory fibers** result in a **loss** of the sensory modality or modalities carried by fibers in that nerve (**anesthesia**).
- **Destructive lesions of motor fibers** result in a paralysis of denervated skeletal muscles. Denervated skeletal muscle fibers exhibit fasciculations (random twitches seen beneath the skin) and may undergo atrophy.

V. **The circulatory system produces, absorbs, and circulates components of the extracellular fluid to maintain cells in all tissues of the body. Cell maintenance includes exchange of nutrients, waste products, and gases.**

   A. The **cardiovascular system** consists of the heart and blood vessels, which begin or end at the level of endothelium-lined capillaries.

   1. Two elastic arteries—the aorta and pulmonary trunk—carry blood from the heart. **Muscular arteries** distribute blood to arterioles, and **arterioles** deliver blood to capillaries where exchange occurs.

   2. Most small arteries communicate with other small arteries through anastomoses.

      a. **Anastomoses** permit collateral circulation where increased blood flow through an anastomosis may compensate for a blockage. Most tissues are supplied by functional end arteries, which have potential anastomotic vessels.

      b. Arteries that lack collateral circulation are **anatomic end arteries.** Examples are few and include the central artery of the retina and the interlobular arteries of the kidney.

   3. **Caval and pulmonary veins** receive blood that has traversed 1 venous capillary bed before returning to the heart.

   4. **Portal veins** receive blood that has traversed 1 venous capillary bed but deliver that blood to a second capillary bed before the blood is returned to the heart. Examples of portal systems are the hepatic portal system and the hypophyseal portal system.

   5. **Arteriovenous shunts** are interconnections between small arteries and veins found in many tissues. Shunts bypass capillary beds and have thick smooth muscle walls.

      a. **Shunts in skin** are important for thermoregulation and are used in an attempt to prevent frostbite.

      b. **Shunts in the walls of the intestines** divert blood into the hepatic portal system when nutrients are not being absorbed.

RAYNAUD'S DISEASE

Raynaud's disease is characterized by arteriovenous shunts that show hyperactive vasoconstriction, resulting in a cooling of the hands and feet. **Cyanosis** may be present in the fingers and toes.

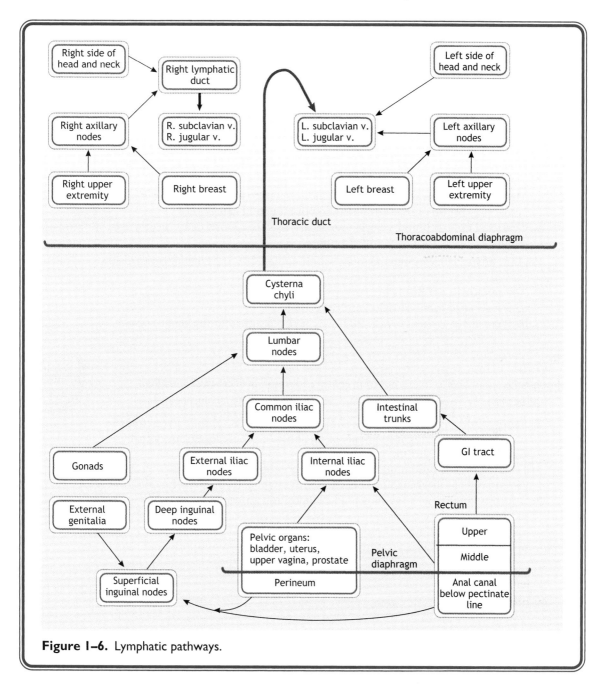

Figure I–6. Lymphatic pathways.

B. The **lymphatic system** drains a component of extracellular fluid containing large proteins back to the venous system from most tissues. Large fatty acids are absorbed from the intestine in lymphatic vessels called **lacteals.**

1. **Lymph capillaries** drain into lymph vessels, which drain into lymph nodes and then into lymph trunks.

2. All **lymph enters the venous system** at the junction of the internal jugular vein and the subclavian vein in the neck by way of the right lymphatic duct on the right or the thoracic duct on the left (Figure 1–6).

3. **Lymph nodes filter and process lymph** by removing debris and trap antigens, foreign material, and malignant cells for exposure to lymphocytes for immune responses.

4. **Metastasizing cancer cells** and infections spread via the lymphatic system. Nodes may provide sites for proliferation or growth of cancer cells.

5. **Tissues that lack blood vessels also lack lymphatic vessels** and include epidermis, cartilage, the CNS, and thymus.

C. **Regions of specialized circulation** are found in the CNS, the pleural, pericardial, and peritoneal cavities, the inner ear, and in the eye.

1. The **cerebrospinal fluid** circulates in and around the CNS.

a. It is secreted into ventricles inside the brain.

b. It circulates in the subarachnoid space and is returned to the venous system.

HYDROCEPHALUS

- **Hydrocephalus** results from an **overproduction or lack of absorption of cerebrospinal fluid.**
- It may also be caused by a blockage inside the ventricles of the CNS or in the subarachnoid space.

2. **Serous fluid** in the pleural, pericardial, and peritoneal cavities serves to promote friction-reducing movements of the lungs, heart, and gastrointestinal viscera below the diaphragm.

3. **Endolymph** is a fluid produced in the inner ear that is characterized by atypical extracellular inorganic ion concentrations essential for hair cell function.

4. **Aqueous humor**, a fluid similar in composition to cerebrospinal fluid, provides nutrients for the lens and the cornea of the eye.

a. Circulation of aqueous humor determines intraocular pressure.

b. Elevated intraocular pressure may result in optic nerve compression, leading to glaucoma.

CLINICAL PROBLEMS

Autonomic Nervous System Match. Choose the best answer from the following choices that best applies to the statement.

A. Sympathetic nervous system

B. Parasympathetic nervous system

C. Both A and B

D. Neither A or B

1. Makes the detrusor muscle contract. B

2. Sends preganglionic axons to synapse in prevertebral ganglia. A

3. Contains preganglionic axons that exit the CNS in ventral roots of spinal nerves. C

4. Has preganglionic neuron cell bodies in the spinal cord. C

5. Sends preganglionic axons to synapse in terminal ganglia. B

6. The superior cervical ganglion is a ganglion in this system. A

7. Facilitates emission. A

8. Postganglionic fibers of this system innervate smooth muscle in blood vessels in the upper limb. A

9. Responsible for innervating a smooth muscle that dilates the pupil. A

10. Increases secretory activity of glands in the wall of the GI tract. B

11. Functions in dilating vascular smooth muscle in erection. B

12. Its preganglionic axons course in white rami. A

13. Its preganglionic axons exit the CNS in cranial nerves V, VII, IX and X. D

14. Its postganglionic axons course in gray rami. A

15. A lesion of this system results in miosis, ptosis, and anhydrosis. A

A 34-year-old woman comes to the physician complaining of blurry vision and droopy eyelids that are more evident late in the day. She talks with a nasal tone to her speech and states that she has difficulty swallowing. Your examination reveals weakness in the ability to adduct the eye in either direction during horizontal gaze; weakness in abduction of the eyes is less pronounced. She has trouble closing her eyes tightly and weakness in pursing her lips and baring her teeth. Sensation and pupillary light reflexes are normal.

16. The patient's signs and symptoms indicate that she may have:
    A. Guillain-Barré syndrome
    B. Shy-Drager syndrome (blurred vision, impotence, dizziness, inability to sweat, urine retention, etc.)
    C. Lambert Eaton syndrome
    D. Multiple sclerosis
    E. Myasthenia gravis (weakness w/ use, ptosis, diplopia, dysphagia, dysarthria, chewing weakness, etc.)

A newborn infant exhibits an inability to pass fecal matter, and imaging reveals a distended colon. Urinary output appears to be normal.

17. Neurons in which of the following may not have developed properly?
    A. Pelvic splanchnic nerves
    B. Terminal ganglia in the wall of the sigmoid colon or rectum
    C. Prevertebral ganglia in the pelvis
    D. Dorsal root ganglia of the sacral spinal nerves
    E. The abdominal part of the sympathetic trunk

Your patient has impotence, urine retention with increased urinary frequency at night, orthostatic hypotension, and an inability to sweat in the upper and lower limbs.

18. Collectively, these symptoms suggest that the patient may have:

    A. Raynaud's disease

    B. Horner's syndrome

    C. Guillain-Barré syndrome

    D. Multiple sclerosis

    E. Shy-Drager syndrome

A patient suffers a fractured rib that lesions the T10 intercostal nerve.

19. The neuronal cell bodies of the lesioned fibers are found in all of the following locations except:

    A. In the dorsal root ganglion of T10

    B. In the T10 paravertebral ganglion

    C. Inside the spinal cord

    D. In a prevertebral ganglion

    E. In a ganglion derived from neural crest cells

ANSWERS

  1. B

  2. A

  3. C

  4. C

  5. B

  6. A

  7. A

  8. A

  9. A

10. B

11. B

12. A

13. D

14. A

15. A

16. The answer is E. The patient exhibits weak muscles innervated almost exclusively by cranial nerves and include CNs III and VI (bilateral ptosis and horizontal diplopia), X (dysphagia and palate droop), and VII (weakness in the muscles of facial expression). The light reflex is intact because the disease spares smooth muscle.

17. The answer is B. The infant most likely has Hirschsprung's disease, which results in an absence of terminal parasympathetic ganglia in the wall of the descending colon, sigmoid colon, or rectum. As a result, there is no peristalsis in the affected segment and a distended bowel proximal to that segment.

18. The answer is E. Shy-Drager syndrome, or multisystem atrophy, results from degeneration of preganglionic sympathetic and parasympathetic neurons in the brainstem and spinal cord and degeneration of neurons in most ganglia. The syndrome is the only choice with symptoms of both sympathetic and parasympathetic divisions.

19. The answer is D. Axons from prevertebral ganglion innervate abdominal pelvic viscera below the diaphragm, and their axons would not be found in the dorsal or ventral rami including an intercostal nerve.

CHAPTER 2
BACK

I. The **vertebral column** contains and protects the spinal cord, supports the skull and the upper limb, and transmits weight to the lower limb through the pelvis.

   A. The vertebral column consists of approximately **33 vertebrae**. There are **24 individual vertebrae: 7 cervical, 12 thoracic, and 5 lumbar.** Five sacral vertebrae are fused to form the sacrum, and typically four coccygeal vertebrae are fused to form the coccyx (Figure 2–1).

   B. Most of the 24 **individual vertebrae** are composed of a body, a vertebral arch, and a pair of costal processes.

   1. The **bodies of vertebrae** are situated anteriorly and increase in size from cervical to lumbar regions.

   2. The **vertebral arch** consists of two pedicles and two laminae. The pedicles attach to the body, and the laminae unite to form the spinous process (Figure 2–2).

SPINA BIFIDA

Spina bifida results when the **laminae fail to fuse** to form a spinous process and is most commonly seen at lower lumbar or sacral vertebral levels.

- In **spina bifida occulta,** one or more spinous processes fail to form at lumbar or sacral levels. This condition is asymptomatic and may be marked by a tuft of hair in skin over the defect.

- In **spina bifida cystica,** a cyst protrudes through the defect in the vertebral arch. These conditions can be diagnosed in utero on the basis of elevated levels of alpha-fetoprotein after amniocentesis and by ultrasound imaging. Spina bifida cystica may result in hydrocephalus and neurological deficits.
  –In **spina bifida cystica with meningocele,** the cyst is lined by the dura and arachnoid and contains cerebrospinal fluid (CSF).
  –In **spina bifida cystica with meningomyelocele,** the lumbosacral spinal cord is displaced into the cyst. Displacement of the cord stretches lumbosacral spinal nerves and may result in bladder, bowel, or lower limb weakness.
  –In **spina bifida with myeloschisis or rachischisis**, the caudal end of the neural tube fails to close in the dorsal midline and is exposed on the surface of the back.

   3. The **costal processes** usually form transverse processes that project laterally at the junction between each lamina and pedicle.

   C. The individual vertebrae and the sacrum have **four curvatures in adults: cervical, thoracic, lumbar, and sacral**. The cervical and lumbar curvatures are the secondary curvatures that are directed anteriorly; the thoracic and sacral curvatures are primary curvatures directed posteriorly.

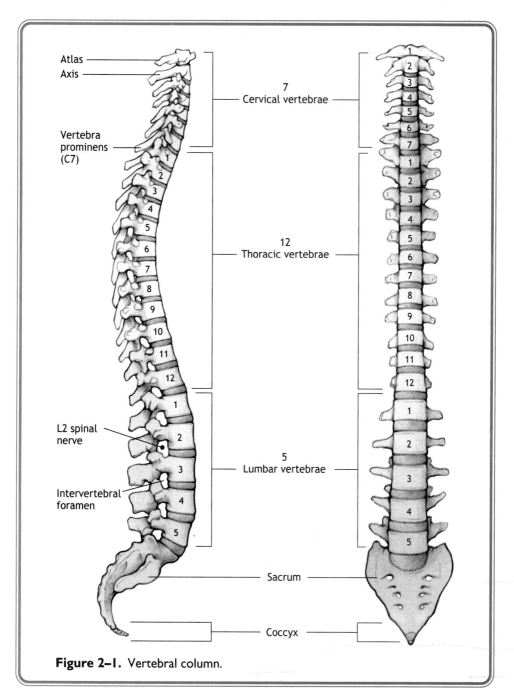

Atlas

Axis

Vertebra prominens (C7)

7
Cervical vertebrae

12
Thoracic vertebrae

L2 spinal nerve

5
Lumbar vertebrae

Intervertebral foramen

Sacrum

Coccyx

Figure 2–1. Vertebral column.

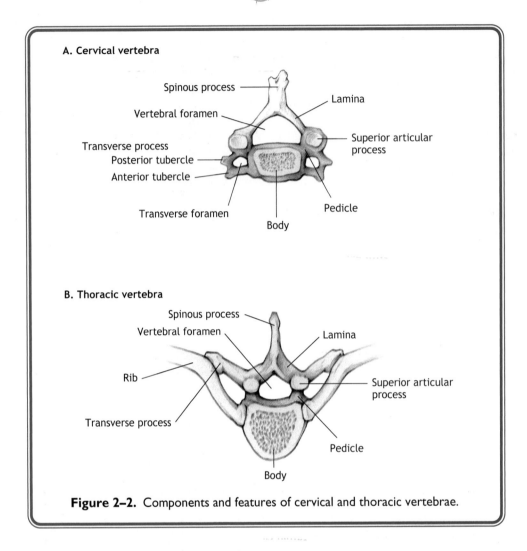

A. Cervical vertebra

Spinous process

Vertebral foramen

Lamina

Superior articular
process

Transverse process
Posterior tubercle
Anterior tubercle

Transverse foramen

Pedicle

Body

B. Thoracic vertebra

Spinous process

Vertebral foramen

Lamina

Rib

Superior articular
process

Transverse process

Pedicle

Body

Figure 2–2. Components and features of cervical and thoracic vertebrae.

ABNORMAL CURVATURES OF THE SPINE

CLINICAL
CORRELATION

- A **kyphosis** is an abnormal increase in the posterior curvature of the spine. The most common kyphosis is a postural kyphosis. Kyphosis may also be caused by anterior wedge-shaped thoracic vertebrae or by resorption of the anterior parts of the thoracic vertebral bodies from osteoporosis.
- A **lordosis** is an abnormal increase in the anterior curvature of the spine that may be caused by a weakening of the anterior abdominal wall musculature as a result of weight gain.
- A **scoliosis** is an abnormal lateral curvature that may be caused by an absent half of a vertebra or a wedge-shaped vertebra or by an asymmetric weakness of back musculature.

    D. Individual vertebrae articulate at facet joints and through intervertebral disks and are supported by ligaments.

        1. Zygapophyseal, or facet, joints are formed between the facets of superior and inferior processes at the junction of each pedicle and lamina. Facet joints permit gliding motions. Intrinsic or deep back muscles act at these joints.

2. An **intervertebral disk** is situated between the bodies of most adjacent vertebrae.
   a. The disk consists of a fibrocartilaginous **annulus fibrosus** that surrounds a **nucleus pulposus**; the nucleus pulposus is the postnatal remnant of the fetal notochord.
   b. Movements of the adjacent vertebral bodies produce compression and tension in different parts of the disk.
   c. The disks **function to absorb shock** and distribute weight over the entire surface of the vertebral bodies.
3. The **anterior longitudinal ligament** covers the anterolateral parts of the vertebral bodies and the disks and functions to limit vertebral extension.
4. The **posterior longitudinal ligament** covers the posterior parts of the vertebral bodies and the disks and functions to limit vertebral flexion. The posterior longitudinal ligament is narrower and weaker than the anterior longitudinal ligament.
5. An **elastic ligamentum flavum** extends between the laminae of adjacent vertebrae. These ligaments also function to limit vertebral flexion and help maintain normal vertebral curvatures.

E. The individual vertebrae are separated by **intervertebral foramina.**
   1. Intervertebral foramina **transmit dorsal and ventral roots of spinal nerves** into and out of the vertebral canal.
   2. Intervertebral foramina are bounded by the pedicles of adjacent vertebrae, posteriorly by the facet joints, and anteriorly by the bodies and the intervertebral disks.

II. The **vertebral canal** is formed by all of the individual vertebral foramina and the ligaments and disks that interconnect them. The vertebral canal contains the meninges, the spinal cord, and the roots of spinal nerves (Figure 2–3).

A. The **epidural space** is outside the dural layer of the meninges and contains fat and the internal vertebral venous plexus.
   1. The **internal venous plexus** connects veins that drain the thorax, abdomen, and pelvis with dural venous sinuses of the cranial cavity.
   2. The internal venous plexus also provides a **route for metastasis** of neoplasms of the prostate, uterus, and rectum to the cranial cavity.

B. The **dura mater** is continuous with the meningeal dura of the cranial cavity and ends at the level of the S2 vertebra. Lateral extensions of the dural sac contain the roots of spinal nerves.

C. The **subdural space** is a potential space between the dura and the arachnoid mater.

D. The **arachnoid mater** also extends to the level of the S2 vertebra and is pressed against the dura by the pressure of CSF.

E. The **subarachnoid space contains CSF**.
   1. CSF is secreted into ventricles of the brain, circulates through the subarachnoid space, and is returned to the venous system.
   2. CSF has a normal pressure of approximately 100 mm H_2O.
   3. CSF cushions the brain and spinal cord, absorbs waste products, and transports hormones.

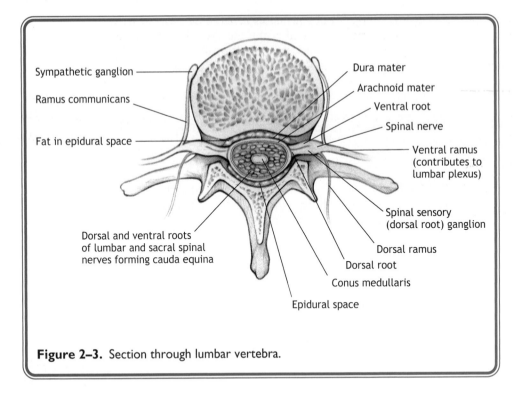

Sympathetic ganglion

Ramus communicans

Fat in epidural space

Dorsal and ventral roots
of lumbar and sacral spinal
nerves forming cauda equina

Dura mater

Arachnoid mater

Ventral root

Spinal nerve

Ventral ramus
(contributes to
lumbar plexus)

Spinal sensory
(dorsal root) ganglion

Dorsal ramus

Dorsal root

Conus medullaris

Epidural space

Figure 2–3. Section through lumbar vertebra.

4. CSF is a clear, colorless fluid containing few cells, a low protein content, and a lower glucose concentration than serum.

F. The **pia mater** covers the spinal cord and the roots of spinal nerves.
1. **Denticulate ligaments** are lateral extensions of the pia that anchor the spinal cord to the dura.
2. The **filum terminale** consists of pia that extends from the inferior end of the cord at the level of the L2 vertebra and joins the dura and arachnoid to end in the sacral canal at the level of the S2 vertebra.

G. The **spinal cord** occupies the superior two-thirds of the vertebral canal and ends inferiorly approximately at the level of the L2 vertebra. The **conus medullaris** is the tapered inferior end of the spinal cord that contains the sacral and coccygeal cord segments and is located at the level of the L2 vertebra.

H. The **spinal nerves** exit the vertebral canal through intervertebral foramina, sacral foramina, or the sacral hiatus.
1. The **first 7 cervical nerves exit superior to the cervical vertebra** for which they are named (eg, the C4 nerve exits between the C3 and the C4 vertebrae).
2. **Beginning with the T1 spinal nerve, all other spinal nerves exit inferior to the vertebra** for which they are named (eg, the L4 nerve exits between the L4 and the L5 vertebrae).
3. The **eighth cervical nerve** exits through the intervertebral foramen between the C7 and T1 vertebrae because there are 8 cervical spinal nerves but only 7 cervical vertebrae.

4. The **cauda equina** is formed by dorsal and ventral roots of lumbar and sacral spinal nerves that extend inferior to the end of the spinal cord at the L2 vertebra.

LUMBAR PUNCTURE

Lumbar puncture is a procedure used to **sample CSF** or to **introduce anesthetic agents** into the subarachnoid space.

- *Lumbar puncture is **typically performed between the L4 and the L5 vertebrae** well below the inferior end of the spinal cord.*
- ***In a midline lumbar puncture a needle will traverse***
  –*Skin*
  –*Superficial and deep fascia*
  –*Supraspinous and interspinous ligaments*
  –*Intralaminar space*
  –*Epidural space*
  –*Dura*
  –*Arachnoid*
- *In a **lumbar puncture off the midline**, the needle will traverse a ligamentum flavum instead of the supraspinous and interspinous ligaments and the intralaminar space.*

RADICULOPATHIES

Radiculopathies result from **compression of the roots of spinal nerves** in the intervertebral foramina or in the vertebral canal.

- *Typical symptoms are **pain and paresthesias** (altered sensations usually in the form of numbness or tingling) in the dermatomes supplied by the compressed sensory roots. The pain may radiate over the dermatomal distribution of the affected sensory roots.* —motor nerve distribution
- *Patients with radiculopathies may also have **weakness of skeletal muscles** in myotomes supplied by the compressed motor roots.*

Radiculopathies may be caused by osteoarthritis, spondylitis, spondylosis, or a herniated disk.

- ***Osteoarthritis** is an inflammation that results in additional bone growth by osteophytes at the facet joints.*
- ***Spondylitis** is an inflammation that results in additional bone growth by osteophytes at the margins of the vertebral bodies. The anterior longitudinal ligament and the sacroiliac joints may undergo calcification. Patients exhibit ankylosis (joint stiffening) and a "bamboo spine" (Marie-Strümpell disease).*
- ***Spondylosis** involves degenerative changes in intervertebral disks, and is usually combined with osteoarthritis at the margins of the vertebral bodies.*

III. Regional characteristics of vertebrae are varied (Figure 2–2).
   A. Cervical Vertebrae
   1. Cervical vertebrae are composed of small bodies, short spinous processes with bifid tips, and transverse processes that transmit the vertebral arteries.
   2. They have **facet joints** of the C3 through C7 vertebrae that are oriented at a 45-degree angle relative to the transverse plane; this alignment permits flexion, extension, lateral bending, and rotation.

WHIPLASH

Whiplash injuries cause the cervical vertebrae to be strongly extended and then strongly flexed and may result in an *anterior dislocation of the facet joints.*

3. Cervical vertebrae have **uncinate processes** on the bodies of the C3-7 vertebrae that form uncovertebral synovial joints with the vertebral bodies superior to them. Osteoarthritic changes in these joints may result in compression of the roots of C3-7 cervical spinal nerves.

4. They include the C1 vertebra, the **atlas**, which has a posterior arch and an anterior arch and no body or spinous process.
   a. There is no intervertebral disk between the C1 and the C2 vertebrae.
   b. The superior articular processes of the atlas articulate with the occipital condyles of the skull, forming the **"yes" joints**; these facet joints permit mainly flexion and extension.

5. They include the C2 vertebra, the **axis**. The dens of the axis articulates with the anterior arch of the atlas, and forms the **"no" or pivot joint,** that permits rotation of the atlas and the skull.
   a. The **dens** is held in place by the transverse ligament of the atlas. Rupture of this ligament causes dislocation of the atlantoaxial joint and displacement of the dens posteriorly into the cervical spinal cord. **Compression** of the ventrolateral part of the cervical spinal cord by the dens may result in **quadriplegia.**
   b. The dens is attached to the margins of the foramen magnum by **alar ligaments. Rupture of alar ligaments** causes excessive rotation of the skull.

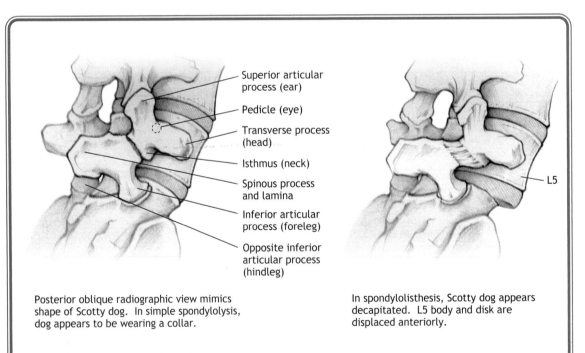

Superior articular
process (ear)

Pedicle (eye)

Transverse process
(head)

Isthmus (neck)

Spinous process
and lamina

Inferior articular
process (foreleg)

Opposite inferior
articular process
(hindleg)

L5

Posterior oblique radiographic view mimics shape of Scotty dog. In simple spondylolysis, dog appears to be wearing a collar.

In spondylolisthesis, Scotty dog appears decapitated. L5 body and disk are displaced anteriorly.

Figure 2–4. Spondylolysis and spondylolisthesis of the L5 vertebra.

6. The **C7 vertebra (vertebra prominens)** has a long spinous process and small transverse foramina that do not transmit the vertebral arteries.

Vertebral arteries only go through C1–C6

HERNIATED NUCLEUS PULPOSUS

*A **herniated disk** occurs when a nucleus pulposus protrudes at the posterolateral part of an annulus fibrosus resulting in compression of roots of lower cervical or lower lumbar spinal nerves. The compressed roots are most commonly those of the more inferior spinal nerve (eg, herniations at the C5-6 disk compress the C6 roots, and herniations at the L4-5 disk compress the L5 roots).*

*A **herniated nucleus pulposus** at **cervical levels** occurs most commonly at the disk between the C6 and C7 vertebrae, which compresses the C7 spinal nerve, or less commonly at the disk between the C7 and the T1 vertebrae, which compresses the C8 spinal nerve.*

- *Compression of the C7 spinal nerve** may result in referred pain in the neck and shoulder and pain and paresthesias in the index and middle fingers. There may be a diminished triceps reflex and weakness in extension of the forearm at the elbow (triceps) or weakness in extension of the wrist and fingers (posterior forearm muscles).*
- *Compression of the C8 spinal nerve** may result in pain in the neck and shoulder and pain and paresthesias in the ring and little fingers. There may be weakness in the hypothenar and interosseous muscles of the hand.*

CERVICAL RIB

*A **cervical rib** may arise from the costal process of C7. The T1 spinal nerve and the subclavian artery may be compressed as they course superior to the cervical rib instead of superior to the first thoracic rib. A patient may present with a diminished radial pulse and pain and paresthesias in the medial forearm. Signs of **Horner's syndrome** may also be seen.*

B. Thoracic Vertebrae

1. Thoracic vertebrae have heart-shaped bodies, long, obliquely oriented, spinous processes, and costal facets on the bodies and on the transverse processes for articulation with ribs.

2. They also have facet joints that are oriented at a 60-degree angle relative to the transverse plane; this alignment permits mainly lateral bending and rotation. Flexion and extension are limited by the fixation provided by the ribs.

C. Lumbar Vertebrae

1. Lumbar vertebrae have large kidney-shaped bodies, short, horizontally oriented spinous processes, and long transverse processes.

2. Their facet joints are oriented perpendicular to the transverse plane; this alignment permits mainly flexion and extension and lateral bending. Rotation is limited.

3. The superior and inferior articular processes are interconnected by an **isthmus or pars interarticularis**; these structures, combined with the spinous process and a transverse process, are in the shape of a **"Scottie dog."**

SPONDYLOLYSIS AND SPONDYLOLISTHESIS

- *In **spondylolysis**, there is a defect or fracture of the isthmus, with no anterior displacement of the vertebral body. Radiographs show that the Scottish terrier appears to be wearing a collar at the site of the fracture (see Figure 2–4).*

- In **spondylolisthesis**, a unilateral or bilateral defect or fracture of the isthmus is accompanied by anterior displacement of the vertebral body (see Figure 2–4). Radiographs show that the head of the "Scottie dog" (the transverse process) appears to be separated from the body. Spondylolisthesis is most common between the L5 vertebra and the sacrum and may stretch roots of lumbosacral spinal nerves in the cauda equina. Patients have bilateral lower back pain that radiates into both lower limbs and weakness in muscles of the legs.
- **Spinal stenosis**, a narrowing of the vertebral canal, can be caused by spondylosis, in which degenerative changes occur in the L4 or the L5 intervertebral disks or by osteoarthritis at the facet joints at these levels.
- A **herniated nucleus pulposus at lumbar levels** occurs most commonly in disks between the L4 and L5 vertebrae or between the L5 and S1 vertebrae (Figure 2–5).
- Small herniations compress the roots of spinal nerves exiting through intervertebral foramina immediately below the affected disk; L4-5 herniations compress the L5 nerve roots, and L5-S1 herniations compress the S1 nerve roots.
- **Compression of the L5 or the S1 spinal nerve roots** may result in sciatica, characterized by pain that radiates from the back into the thigh, leg, and foot.
  –Compression of L5 may result in pain and paresthesias in the posterior thigh, the anterolateral leg, and dorsum of the foot. There may be weakness in extension of the great toe (extensor hallucis longus) and weakness in dorsiflexion (tibialis anterior).
  –**Compression of the S1 spinal nerve roots** may result in pain and paresthesias in the posterolateral leg, heel, and lateral side of the foot. There may be weakness in flexion of the leg at the knee (hamstrings), weakness in plantar flexion (gastrocnemius and soleus), and a diminished Achilles tendon reflex.

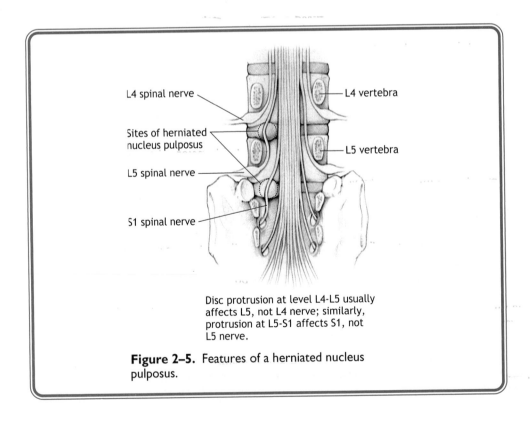

Figure 2–5. Features of a herniated nucleus pulposus.

D. **The Sacrum**
1. The **sacrum** is composed of **five fused sacral vertebrae**. The superior aspect of the sacrum bears and transmits weight to the pelvic girdle.
2. It contains a **median crest**, which represents the fused sacral spinous processes, and the intermediate crests, which represent the fused articular processes.
3. The **promontory** forms the ventral surface of the S1 vertebra, and is a boundary of the pelvic inlet important in obstetrics.
4. The sacrum articulates at the sacroiliac joints, which are formed by the lateral surfaces of the superior part of the sacrum and the medial surface of each ilium.
5. It contains four pairs of **dorsal sacral foramina** and four pairs of **ventral sacral foramina** that transmit the dorsal rami and the ventral rami of the S1 through the S4 sacral spinal nerves, respectively.
6. The sacrum contains the sacral canal, which ends at the sacral hiatus in the dorsal midline; the canal contains the roots of the S1-coccygeal spinal nerves, and the sacral hiatus transmits the S5 and coccygeal spinal nerves.

EPIDURAL OR CAUDAL BLOCK

An **epidural or caudal block** is performed by administering anesthetic through the sacral hiatus, which diffuses through the meninges and anesthetizes the roots of the sacral and coccygeal spinal nerves in the cauda equina.

E. **The Coccyx**
1. The **coccyx** is formed by **three to five fused coccygeal vertebrae.**
2. It is an attachment site for the gluteus maximus and for the anococcygeal ligament, which is an attachment site for muscles of the pelvic diaphragm.

IV. The **muscles of the back** are divided into three groups based on their function, their innervation, and their attachments to bony structures (Table 2–1).
A. **Superficial Muscles of the Back**
1. The **superficial muscles of the back** are attached to the pectoral girdle and act on the upper extremity.
2. They include the trapezius, latissimus dorsi, rhomboid major, rhomboid minor, and levator scapulae.
3. These muscles are innervated by ventral rami of spinal nerves through branches of the brachial plexus, except for the trapezius, which is supplied by CN XI, the accessory nerve.

B. **Intermediate Muscles of the Back**
1. The **intermediate muscles of the back** attach to the ribs and act as accessory muscles of respiration.
2. They include the serratus posterior superior, serratus posterior inferior, and 12 pairs of levator costarum muscles.
3. The intermediate muscles of the back are innervated by ventral rami of spinal nerves.

C. **Deep or Intrinsic Muscles of the Back**
1. The **deep or intrinsic muscles of the back** attach mainly to transverse and spinous processes of vertebrae and act on the vertebral column at the intervertebral joints.

Table 2–1. Actions of back muscles.

Action	Muscles Involved	Innervation	Major Segments of Innervation
Superficial layer: act on pectoral girdle on humerus	Trapezius	Accessory (CN XI)	
	Latissimus dorsi	Thoracodorsal	C6, C7, C8
	Levator scapulae	Dorsal scapular	C5
	Rhomboideus major	Dorsal scapular	C5
	Rhomboideus minor	Dorsal scapular	C5
Intermediate layer: Act as accessory muscles of respiration (attach to ribs and vertebrae)	Serratus posterior superior	Ventral rami of spinal nerves	
	Serratus posterior inferior	Ventral rami of spinal nerves	
	Levatores costarum (12 pairs)	Ventral rami of spinal nerves	
Deep layers: act on vertebral column	1. Splenius capitus, cervicis	Dorsal rami of spinal nerves	
	2. Erector spinae: Iliocostalis Longissimus Spinalis	Dorsal rami of spinal nerves	
	3. Transversospinalis: Semispinalis Multifidus Rotatores	Dorsal rami of spinal nerves	
Suboccipital: act on altas and axis	Rectus capitis posterior minor	Dorsal ramus of CI (suboccipital n.)	CI
	Rectus capitis posterior major		CI
	Inferior oblique (capitis)		CI
	Superior oblique (capitis)		CI

CN, cranial nerve.

2. They consist of the **erector spinae**, which includes three parallel muscle groups: from lateral to medial, the iliocostalis, the longissimus, and the spinalis.
   a. Acting bilaterally, the components of the erector spinae extend the vertebral column at intervertebral joints.
   b. Acting unilaterally, the components of the erector spinae produce lateral bending of the vertebral column at intervertebral joints.

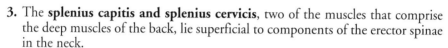

3. The **splenius capitis and splenius cervicis**, two of the muscles that comprise the deep muscles of the back, lie superficial to components of the erector spinae in the neck.
   a. The splenius capitis acts to extend the head; the splenius cervicis acts to rotate the head.
   b. Acting unilaterally, both muscles produce lateral bending of the cervical vertebrae.
4. The deep muscles of the back also consist of the **transversospinalis muscles**, which include three muscle groups that lie deep to the erector spinae: from superficial to deep, the semispinalis, multifidus, and rotatores.
   a. Acting bilaterally, these muscles act to extend the vertebral column.
   b. Acting unilaterally, these muscles produce rotation of the vertebral column.
5. The muscles of the **suboccipital triangle** contribute to extension at the atlantooccipital joints and rotation at the atlantoaxial joints.
6. The deep muscles of the back are innervated by **dorsal rami of spinal nerves**.

CLINICAL PROBLEMS

A 45-year-old man complains of low back pain that radiates into both lower limbs and leg weakness. A magnetic resonance image (MRI) scan reveals that the body of the L5 vertebra is displaced anteriorly.

1. A diagnosis of the patient's condition would be:
   A. Spondylolysis
   B. Spondylosis
   C. Spondylitis
   D. Spondylolisthesis
   E. Spinal stenosis

A spinal anesthesia is used for a patient undergoing surgery. A lumbar puncture is performed in the midline between the L4 and L5 vertebrae.

2. Which of the following structures will not be traversed by the needle in a normal performance of the procedure?
   A. Interspinous ligament
   B. Posterior longitudinal ligament
   C. Epidural space
   D. Arachnoid
   E. Dura

3. Which of the following structures in the vertebral canal will be anesthetized in a spinal procedure?
   A. Dorsal and ventral rami

B. Lumbar spinal nerves

C. Sacral spinal nerves

D. Dorsal and ventral roots

E. Lumbosacral plexus

An MRI reveals that a patient has a posterolateral herniation of the nucleus pulposus of the intervertebral disk between L5 and the sacrum.

4. Which of the following might you expect to observe in the patient?

A. Weakness in dorsiflexion

B. Altered sensation on the dorsum of the foot

C. Weakness in plantar flexion

D. Altered sensation in the anterior thigh

E. Weakness in extension of the leg at the knee

A 50-year-old man complains of back pain and has difficulty walking. Diagnostic imaging reveals calcification of the sacroiliac joints and anterior longitudinal ligament resulting from additional bone growth by osteophytes.

5. These findings suggest that the patient has

A. Spondylitis

B. Spondylosis

C. Spondylolysis

D. Spinal stenosis

E. Scoliosis

A 64-year-old man presents with pain that radiates from the back, through the posterior thigh, and into the leg and foot. The diagnosis is a herniated nucleus pulposus of the intervertebral disk between the L4 and the L5 vertebrae.

6. What else might the patient experience?

A. Altered sensation in the L3 dermatome

B. Weakness of muscles innervated by the L5 spinal cord segment

C. Inability to contract and empty the bladder

D. Fecal incontinence

E. Weakness in the ability to extend the leg at the knee

7. A patient develops a peripheral neuropathy that results in the degeneration of nerve fibers in dorsal rami. All of the following may be evident EXCEPT:

A. Sensation may be altered in skin covering the trapezius.

B. The iliocostalis muscle may be weak.

C. Axons exiting through the dorsal sacral foramina may be affected.

D. Motor fibers in the dorsal scapular nerve may be affected.

E. Sensation from facet joints may be altered.

8. Your patient suffers from a herniated disk between the L5 vertebra and the sacrum. If the herniation is small, what neural structure might be subject to compression?

 A. L4 spinal nerve

 B. L5 spinal nerve

 C. S1 spinal nerve

 D. Conus medullaris

 E. Lumbar splanchnic nerve

ANSWERS

1. The answer is D. In *spondylolisthesis*, there is an anterior displacement of a vertebral body relative to the vertebral body inferior to it. Spondylolisthesis is usually accompanied by a defect or fracture of the isthmus of that vertebra. Spondylolisthesis is most common between the L5 vertebra and the sacrum and, in this patient, has stretched roots of lumbosacral spinal nerves in the cauda equina. This results in lower back pain radiating into both lower limbs and weakness in muscles of the legs.

2. The answer is B. The posterior longitudinal ligament covers the posterior parts of the vertebral bodies and the intervening disks and is anterior to the dural sac; the needle pierces the posterior aspect of the sac in a normal performance of a spinal procedure.

3. The answer is D. The dorsal and ventral roots are in the cauda equina, found in the vertebral canal; the other choices are found either in the intervertebral foramen or outside the vertebral canal.

4. The answer is C. The roots of the S1 spinal nerve are typically compressed by a posterolateral herniation of the disk between L5 and S1. Compression of the S1 roots may result in altered sensation in the posterolateral leg, heel, and lateral side of the foot, weakness in flexion of the leg at the knee, and weakness in plantar flexion.

5. The answer is A. In *spondylitis*, additional bone growth by osteophytes at the margins of the vertebral bodies may result in calcification of the anterior longitudinal ligament and the sacroiliac joints.

6. The answer is B. The L5 spinal nerve has been compressed. All other choices reflect lesions to spinal nerves other than L5.

7. The answer is C. The dorsal scapular nerve is a ventral ramus of the C5 spinal nerve. Few dorsal rami have names.

8. The answer is C. S1 spinal nerve. Disk herniations always compress the nerve that corresponds to the more inferior numbered vertebra.

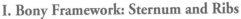

CHAPTER 3
THORAX

I. Bony Framework: Sternum and Ribs

A. The **sternum** consists of the manubrium, the body, and the xiphoid process.

1. The **manubrium** contains the jugular notch, articulates with the medial end of each clavicle at the sternoclavicular joints, and articulates with the costal cartilages of the first two ribs.

2. The **body** articulates with the manubrium at the sternal angle. The costal cartilage of the second rib articulates at the sternal angle, and the costal cartilages of ribs 3 through 7 articulate with the body inferior to the sternal angle.

3. The **xiphoid process** articulates with the body and provides an attachment site for the diaphragm and the abdominal musculature.

4. The **sternal angle** is located at the junction of the manubrium with the body of the sternum.

CLINICAL CORRELATION

THE STERNAL ANGLE

The sternal angle is an important anatomic landmark. It is located at the level of the horizontal plane intersecting the sternal angle and the disk between the T4 and the T5 vertebrae. Here:

- *The trachea bifurcates into the right and left primary bronchi.*
- *The arch of the aorta arises from the ascending aorta and continues as the descending aorta.*
- *The azygos vein drains into the superior vena cava.*

B. **Twelve pairs of ribs** articulate posteriorly with the thoracic vertebrae.

1. The **head** of most ribs articulates with the bodies of adjacent thoracic vertebrae and the intervertebral disk at the costovertebral joints.

2. The **tubercle** of most ribs articulates with a transverse process at a costotransverse joint.

3. **True ribs (ribs 1–7)** have costal cartilages that articulate individually with the body of the sternum. Elevation of the true ribs combined with the anterior movement of the sternum increases the anteroposterior diameter of the thorax during inspiration (a pump handle movement).

4. **False ribs (ribs 8–10)** have costal cartilages that articulate with a more superior costal cartilage and contribute to the costal margin. Elevation of the false ribs increases the transverse diameter of the thorax during inspiration (a bucket handle movement).

5. **Floating ribs (ribs 11 and 12)** do not articulate with the sternum or with the transverse process of thoracic vertebrae. They are embedded in musculature of the abdominal wall.

RIB FRACTURES

Fractures of a rib commonly occur just anterior to the **angle** of the rib, the **weakest point of the rib**, and may cause a **pneumothorax** (see later discussion).

II. Skeletal Muscles of the Thoracic Wall

A. **Skeletal muscles** of the thoracic wall act on the ribs and sternum during respiration.

B. The main components are the diaphragm, the external intercostal muscles, and the internal intercostal and transversus thoracis muscles.

   1. The diaphragm
      a. The **diaphragm** is the main muscle of inspiration.
      b. It acts to elevate the ribs and increase the anteroposterior, transverse, and vertical diameters of the thorax.
      c. The diaphragm is innervated by the **phrenic nerves**, which consist of ventral rami from the C3, C4, and C5 spinal cord segments.
         (1) The phrenic nerves provide the motor and most of the sensory innervation of the diaphragm.
         (2) Sensory fibers in intercostal nerves innervate the periphery of the diaphragm.

PHRENIC NERVE LESIONS

• An **irritative lesion of a phrenic nerve** causes involuntary contractions of the diaphragm and may result in **hiccups**.

• A **destructive lesion of a phrenic nerve** may result in paralysis and **paradoxical movement** of one half of the diaphragm. The paralyzed dome of the diaphragm fails to descend upon inspiration, and is forced superiorly by an increase in intra-abdominal pressure.

   2. The external intercostal muscles (Figure 3–1).
      a. The **external intercostal muscles** elevate the false ribs, which increases the transverse diameter of the thorax.
      b. They also elevate the true ribs, which increases the anteroposterior diameter of the thorax.
   3. The **internal intercostal and transversus thoracis muscles** aid in expiration; passive elastic recoil of the thorax is the main mechanism of expiration.
   4. The **sternocleidomastoid, scalene** (anterior, middle, and posterior), and **pectoralis major and minor muscles** also attach to ribs and act as accessory muscles of inspiration.
   5. **Muscles of the anterior abdominal wall**—the rectus abdominis, the external oblique, and the internal oblique—are used in forced expiration and in the Valsalva maneuver (forced expiration against a closed airway).

III. Innervation and Blood Supply of the Thoracic Wall (Figure 3–1)

A. **Intercostal nerves**—the ventral rami of the T1 through T11 spinal nerves— and the **subcostal nerve**—the ventral ramus of T12—innervate the intercostal muscles and the skin of the thoracic wall.

B. Branches of three arteries—**a pair of internal thoracic arteries** and the **descending aorta**—supply the thoracic wall.

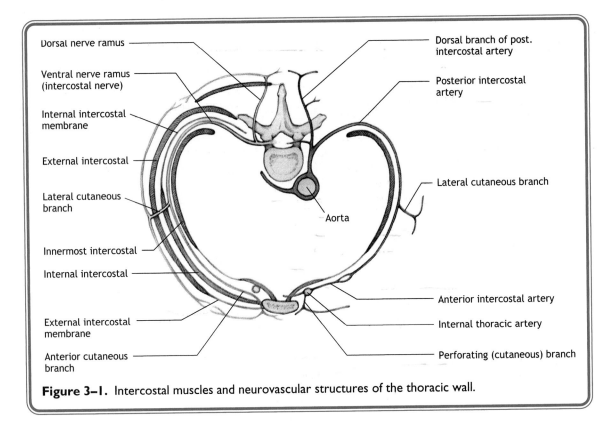

Figure 3–1. Intercostal muscles and neurovascular structures of the thoracic wall.

1. **Anterior intercostal arteries** supply the anterior thoracic wall.
   a. Anterior intercostal arteries arise from the internal thoracic artery or a branch of it, the musculophrenic artery.
   b. The internal thoracic artery is a branch of the subclavian artery.
2. **Posterior intercostal arteries** supply the posterior and lateral thoracic walls and arise mainly from the descending aorta.

C. Tributaries of the **internal thoracic veins** and the **azygos system** of veins drain the thoracic wall.
   1. **Anterior intercostal veins** drain the anterior chest wall and empty into the internal thoracic veins, which drain into the brachiocephalic veins.
   2. **Posterior intercostal veins** drain the posterior and lateral thoracic walls and empty into hemiazygos veins on the left and the azygos vein on the right (Figure 3–2).
      a. The hemiazygos veins cross the midline to drain into the azygos vein.
      b. The azygos vein drains into the superior vena cava.

D. The **intercostal nerves** and the intercostal arteries and veins course through the superior part of each intercostal space protected by the costal margin of an overlying rib. In these neurovascular bundles, the vein lies superior, the artery is in the middle, and the nerve lies inferiorly (a "VAN" arrangement).

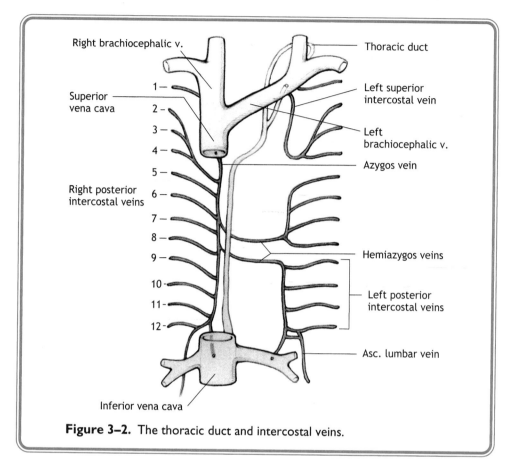

Figure 3–2. The thoracic duct and intercostal veins.

IV. The Breasts

A. **Breasts** are **modified sweat glands** situated in the superficial fascia of the thoracic wall that enlarge in females during menstrual cycles and during pregnancy and produce milk after the birth of an infant.

B. They consist of 15–20 lobes that open by way of individual **lactiferous ducts** on to the nipple. Suspensory (Cooper's) ligaments separate the lobes and attach to the skin.

C. The breasts are supplied mainly by branches of the internal thoracic, lateral thoracic, and anterior and posterior intercostal arteries.

D. They are **divided into 4 quadrants**.
   1. Most of the **lymphatic drainage** from all quadrants of the breast, including the nipple, is to **axillary lymph nodes** (see Figure 1–6).
   2. Parts of the medial quadrants drain to parasternal lymph nodes.

ADENOCARCINOMAS OF THE BREASTS

• Most **breast adenocarcinomas** are lactiferous duct carcinomas that begin as painless masses most commonly in the upper lateral quadrant.

CLINICAL
CORRELATION

- *Late-stage adenocarcinomas* may cause retraction and fixation of the nipple and dimpling of skin, which results from invasion of the suspensory ligaments.
- Breast *adenocarcinomas metastasize* mainly to axillary lymph nodes but also to parasternal nodes, to the opposite breast, and to nodes of the anterior abdominal wall.
- In a radical mastectomy, the breast is removed along with the pectoralis major and minor muscles, axillary lymph nodes and vessels, and tributaries of the axillary vein.
- The **long thoracic nerve** or the **thoracodorsal nerve** may be damaged in a mastectomy procedure. A lesion of the long thoracic nerve may result in weakness in protraction and upward rotation of the scapula and a "winged scapula" at rest. A lesion of the thoracodorsal nerve may result in weakness in extension, adduction, and medial rotation of the arm at the glenohumeral joint.

V. Pleura

A. The **pleura** is one of three thin, friction-reducing, serous membranes in the body; the others are the pericardium and the peritoneum.

B. **Parietal pleura** covers the thoracic wall (costal pleura), diaphragm (diaphragmatic pleura), and mediastinum (mediastinal pleura). Cervical parietal pleura (the cupola) extends into the root of the neck superior to the medial one-third of the clavicle and the first rib (Figure 3–3).

C. **Visceral pleura** completely covers each lung and is continuous with parietal pleura at the root of the lung.

D. The **pleural cavity** is a potential space between the visceral pleura and the parietal pleura.
   1. The pleural cavity contains a thin film of serous fluid, which permits visceral pleura and the underlying lung to glide against the parietal pleura during respiration.
   2. During **inspiration**, skeletal muscles of inspiration contract and create a transient negative pressure in the pleural cavity by increasing the volume of

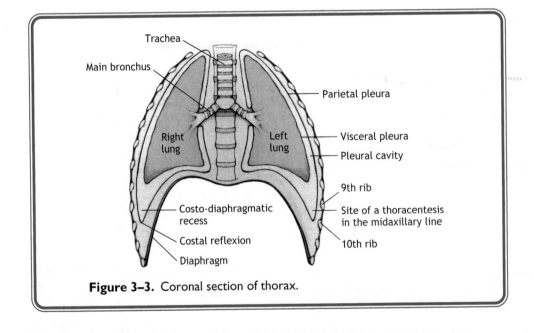

Figure 3–3. Coronal section of thorax.

the thorax. The negative pressure is transferred to the lungs, and the lungs fill with air.

   a. During inspiration, the inferior extent of the visceral pleura and the underlying lung are approximately two rib levels superior to the inferior extent of the parietal pleura.

   b. The inferior extent of visceral pleura and lung is at the level of the 6th rib in the midclavicular line, the 8th rib in the midaxillary line, and the 10th rib in the paravertebral line.

   c. The inferior extent of the parietal pleura is at the level of the 8th rib in the midclavicular line, the 10th rib in the midaxillary line, and the 12th rib in the paravertebral line.

3. The **costodiaphragmatic and costomediastinal recesses** are potential spaces in the pleural cavity at points where the visceral pleura and underlying lung do not completely fill the pleural cavity during inspiration (Figure 3–3).

   a. In the **costodiaphragmatic recess**, the diaphragmatic pleura is in contact with the costal pleura.

   b. In the **costomediastinal recess,** the costal pleura is in contact with the mediastinal pleura.

   c. The costomediastinal recess is filled by the lingula of the superior lobe of the left lung during maximum inspiration.

E. The **costal parietal pleura is innervated by intercostal nerves**, and the mediastinal pleura and most of the diaphragmatic parietal pleura are innervated by the phrenic nerve; the parietal pleura is sensitive to pain.

F. The **visceral pleura is innervated by sensory nerves that course with autonomic nerves**; visceral pleura is insensitive to pain.

PLEURITIS OR PLEURISY

In **pleuritis or pleurisy**, the visceral or parietal pleura is inflamed and becomes rough; adhesions between these two layers may result.

• During respiration, friction created by the adhesions may be audible as a **pleural rub**.

• Patients with **costal pleurisy** may experience sharp pain localized over the adhesion site that increases with inspiration.

• Patients with **mediastinal or diaphragmatic pleurisy** may have pain that is referred over the C3 through C5 dermatomes in the supraclavicular region.

THORACENTESIS

In a **thoracentesis**, a needle is used to **sample or withdraw fluid** from a costodiaphragmatic recess.

• The needle may be introduced into the pleural cavity in the midaxillary line in the ninth intercostal space after passing through skin, superficial fascia, the 3 layers of intercostal muscles, and the parietal pleura.

• To avoid the intercostal nerves, the needle is inserted into the inferior part of the interspace. To anesthetize the intercostal nerve for relief of pain associated with a rib fracture, the needle is inserted into the superior part of the interspace.

PNEUMOTHORAX AND PLEURAL EFFUSIONS

• In a **pneumothorax**, air is introduced into a pleural cavity, and the lung may undergo a partial or complete atelectasis (collapse). Patients with pneumothorax experience pain and difficulty breathing or shortness of breath.

- In an **open pneumothorax**, a penetrating wound of the chest wall pierces the costal pleura, or a penetrating wound in the root of the neck pierces the cervical pleura. Pleuritic pain results from stimulation of the intercostal nerves.
  –Air enters the affected pleural cavity during inspiration, but the negative intrapleural pressure is lost and the lung on the affected side collapses. The heart and other mediastinal structures shift away from the affected side and compress the opposite lung.
  –During expiration, air is expelled from the affected pleural cavity through the wound, and the heart and other mediastinal structures shift back to the affected side. Shifting of mediastinal structures reduces venous return to the heart.
- In a **tension pneumothorax**, a penetrating wound of the pleura creates a valve-like effect in the pleura during respiration.
  –Air enters the pleural cavity during inspiration, and the lung on the affected side collapses; the heart and other mediastinal structures shift away from the affected side and compress the opposite lung.
  –During expiration, a flap of pleural tissue that closes the wound prevents the expulsion of air. With each breath, intrapleural pressure is increased, and the shift of the heart and mediastinal structures to the opposite side is augmented. Cardiac output, venous return, and respiratory function are compromised.
- A **pleural effusion** is an accumulation of fluid in a pleural cavity and may be caused by obstruction of veins or lymphatic vessels that drain the thorax or by an inflammation of structures near the pleura.
- In a **hemothorax**, the blood accumulates in a pleural cavity from hemorrhage of anterior or posterior intercostal vessels or internal thoracic vessels.
- In a **chylothorax**, lymph accumulates in a pleural cavity as a complication of mediastinal surgery or trauma that injures the thoracic duct.

VI. The Lungs

A. The **right lung** is larger than the left and is separated into **superior, middle, and inferior lobes** by an oblique fissure and a horizontal fissure.
   1. The **horizontal fissure** separates the superior and middle lobes.
   2. The **oblique fissure** separates the superior and middle lobes from the inferior lobe.
B. The **left lung** is separated into a superior lobe and an inferior lobe by an oblique fissure.

BREATH SOUNDS

CLINICAL CORRELATION

- **Breath sounds from the superior lobe** of each lung may be auscultated on the anterior and superior aspects of the thoracic wall.
- **Breath sounds from the inferior lobe** of each lung may be auscultated on the posterior and inferior aspects of the back.
- **Breath sounds from the middle lobe** of the right lung may be auscultated on the anterior chest wall near the sternum, inferior to the right fourth costal cartilage.

C. Each lung has a **root** at the site where the lung becomes covered by visceral pleura (Figure 3–4).
   1. Structures in the root enter or exit the lung through the hilus. **Each root contains**
      a. A main bronchus
      b. A pulmonary artery and two pulmonary veins
      c. Bronchial arteries that supply lung tissue
      d. Autonomic nerves, sensory nerves, and lymphatic vessels
   2. In the **hilus of the right lung**, the upper lobar bronchus is superior to the pulmonary artery and the pulmonary veins.

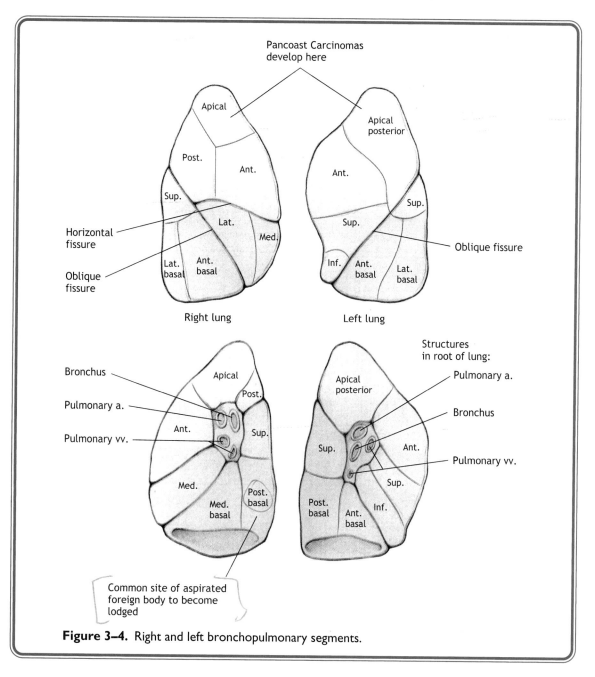

Figure 3–4. Right and left bronchopulmonary segments.

3. In the **hilus of the left lung**, the pulmonary artery is superior to the main bronchus and the pulmonary veins.

FOREIGN BODY ASPIRATION

- An **aspirated foreign body** is more likely to enter the **right main bronchus** because it is shorter, wider, and more vertical than the left main bronchus.
- In a patient who is standing or sitting, the foreign body tends to become lodged in the posterobasal segment of the **inferior lobe of the right lung** (Figure 3–4).

    D. In each lung, the **main bronchus** divides into lobar bronchi (3 on the right, 2 on the left), and each lobar bronchus divides into 10 segmental bronchi that serve restricted areas of lung tissue, the **bronchopulmonary segments** (Figure 3–4).

      1. A segmental branch of a pulmonary artery supplies each segment.

      2. Pulmonary veins are intersegmental and drain parts of adjacent segments.

EMPHYSEMA

- In **emphysema**, respiratory tissue is destroyed, resulting in a permanent abnormal enlargement and increased radiolucency of the affected air spaces and the formation of blebs or bullae.
- In a **spontaneous pneumothorax**, an emphysematous bleb spontaneously ruptures, air is introduced into the pleural cavity through the visceral pleura, and conditions similar to an open or a tension pneumothorax result.
  – The most common site of a spontaneous pneumothorax is in the visceral pleura of the superior lobe of a lung.

    E. The **lymphatic drainage** of each lung is mainly to bronchopulmonary nodes at the hilus, to tracheobronchial nodes and tracheal nodes adjacent to the trachea, and then to a bronchomediastinal trunk.

      1. Lymph from the right lung enters a right bronchomediastinal trunk, which empties into the right lymphatic duct. The right lymphatic duct drains into the junction of the right internal jugular vein and the right subclavian vein.

      2. Most of the **lymph from the left lung** enters a left bronchomediastinal trunk, which empties into the thoracic duct; lymph from the inferior lobe of the left lung drains into the right lung pathway. The thoracic duct drains into the junction of the left internal jugular vein and the left subclavian vein.

BRONCHOGENIC CARCINOMA

Bronchogenic carcinomas may metastasize through lymph channels but may also penetrate the wall of a tributary of a pulmonary vein and metastasize through the pulmonary and the systemic circulations.

- **Supraclavicular lymph nodes** may act as **sentinel nodes** indicating the presence of a malignancy.
- Enlarged supraclavicular lymph nodes on the right may indicate malignancy in the thorax.
- Enlarged supraclavicular lymph nodes on the left may indicate a malignancy in the thorax, abdomen, or pelvis, because all lymph below the diaphragm is returned to the venous system on the left by way of the thoracic duct.

A **Pancoast carcinoma** that develops in the apical part of the superior lobe of either lung may cause thoracic outlet syndrome.

• **Thoracic outlet syndrome** results from compression of the sympathetic trunk at the level of the stellate ganglion, the inferior trunk of the brachial plexus in the root of the neck, the subclavian vessels, or the recurrent laryngeal nerve.

–Compression of the sympathetic trunk or the inferior trunk of the brachial plexus may cause **Horner's syndrome.** Signs and symptoms of Horner's syndrome include anhydrosis, a loss of sweating on corresponding side of the face, ptosis, a drooping of the upper eyelid, and miosis, a constriction of the pupil.

–A **decreased radial pulse** in the upper limb caused by compression of the subclavian artery and vein.

–**Hoarseness and dysphagia** resulting from compression of a recurrent laryngeal nerve.

–Compression of the C8 and T1 ventral rami in the inferior trunk of the brachial plexus may also cause **paresthesia** in the forearm and hand and weakness and atrophy of hand muscles.

Primary carcinomas that develop in other organs commonly metastasize to the lungs.

F. **Sympathetic and parasympathetic nerves** innervate glands and smooth muscle in the trachea, bronchi, and the lungs.
1. Both types of **autonomic fiber** reach respiratory structures by way of pulmonary plexuses, which are extensions of the cardiac plexus.
2. **Each pulmonary plexus** contains:

   – CN XI → accessory nerve

   a. Preganglionic parasympathetic axons from the vagus nerves (CN X)
   b. Postganglionic parasympathetic axons from terminal ganglia in the plexus
   c. Postganglionic sympathetic axons from cervical and upper thoracic (cardiopulmonary) splanchnic nerves
      (1) These splanchnic nerves have their cell bodies in the superior, middle, and inferior cervical paravertebral ganglia and in the T1 through T5 paravertebral ganglia.
      (2) Preganglionic sympathetic axons that synapse in these paravertebral ganglia arise from the T1 through T5 spinal cord segments.
   d. Sensory fibers from baroreceptors in the trachea, bronchi, and lungs
      (1) These fibers course with and enter the central nervous system (CNS) in the vagus nerves.
      (2) The lungs and visceral pleura are insensitive to pain.
3. The **sympathetic innervation** facilitates relaxation of smooth muscle in the bronchial tubes, causing a dilation of the tubes and allowing a greater volume of respiration.
4. The **parasympathetic innervation** facilitates contraction of smooth muscle, which constricts the bronchial tubes.

VII. Mediastinum

A. The mediastinum is the **region between the pleural cavities.**

B. The mediastinum is separated into **two divisions**—superior and inferior—by a horizontal plane passing from the sternal angle through the intervertebral disk between the T4 and the T5 vertebrae (Figure 3–5).
1. The **superior mediastinum** contains structures that lie between the first thoracic vertebra and the first rib and the horizontal plane between the sternal angle and the intervertebral disk between the T4 and the T5 vertebra (Figure 3–6).
   a. The **endocrine layer** contains the thymus in children or remnants of the thymus in adults. The thymus extends inferiorly into the anterior mediastinum in children.

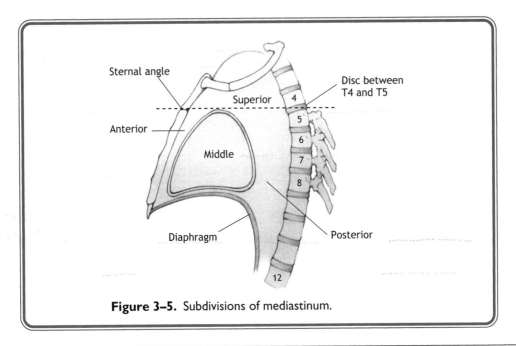

Figure 3–5. Subdivisions of mediastinum.

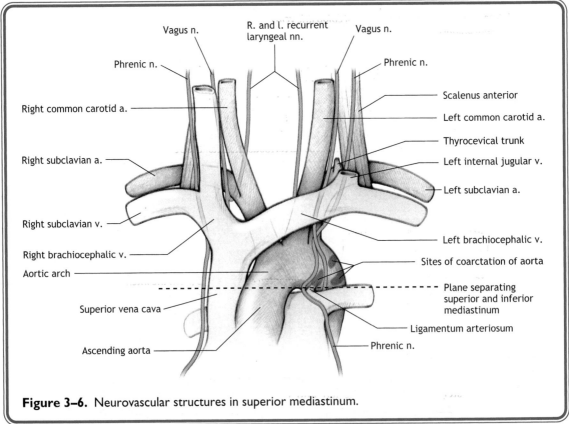

Figure 3–6. Neurovascular structures in superior mediastinum.

THYMOMA

*A **thymoma** may develop in the superior or anterior mediastinum; patients with a thymoma may also have **myasthenia gravis**. Signs and symptoms include an obstructed left brachiocephalic vein and chest pain.*

     b. The **venous layer** contains the superior vena cava and the left and right brachiocephalic veins. The superior vena cava descends inferiorly on the right side of the superior mediastinum and empties into the right atrium.

       (1) Each brachiocephalic vein is formed posterior to a sternoclavicular joint by the union of an internal jugular vein and a subclavian vein.

       (2) The left brachiocephalic vein crosses the midline behind the thymus or thymic remnants and receives inferior thyroid veins and the left superior intercostal vein.

SUPERIOR VENA CAVA SYNDROME

*In **superior vena cava syndrome**, the superior vena cava may be compressed by lymph nodes enlarged because of metastasis from a bronchogenic carcinoma or because of the carcinoma itself.*

- *Patients with superior vena cava syndrome may experience headache, edema of the head and neck, prominent superficial veins, and cyanosis. The veins of the upper limbs fail to empty when the limb is elevated above the heart.*
- *In a complete occlusion of the superior vena cava, venous return from the head, neck, and upper limbs is shunted into tributaries of the inferior vena cava.*
- ***Anastomoses of the superior vena cava** to the inferior vena cava may occur*
   –Between lateral thoracic veins and superficial epigastric veins
   –Between superior epigastric veins and inferior epigastric veins

     c. The **arterial layer** contains the arch of the aorta and its three branches: the brachiocephalic artery, the left common carotid artery, and the left subclavian artery (Figure 3–7a, b). The brachiocephalic artery divides into the right common carotid artery and the right subclavian artery.

       (1) The ligamentum arteriosum, a remnant of the fetal ductus arteriosus, is situated between the left pulmonary artery and the arch of the aorta.

       (2) During fetal life, the ductus arteriosus shunts blood from the pulmonary trunk directly into the arch of the aorta, bypassing the lungs.

COARCTATION OF THE AORTA

*A **coarctation of the aorta** is a **constriction of the aorta** that occurs just **proximal** (infantile type) or **distal** (adult type) to the ligamentum arteriosum.*

- *In these patients, blood pressure is reduced in the lower limbs and elevated in the head, neck, and upper limbs.*
- *Anastomoses in the intercostal spaces between the anterior intercostal arteries (from the internal thoracic artery) and posterior intercostal arteries (from descending aorta) provide collateral circulation that bypasses the coarctation. Blood flows in the retrograde direction through the posterior intercostal arteries into the descending aorta.*
- *Dilation of the anterior and posterior intercostal arteries may result in resorption of ribs and a **"notching"** observable on radiographs.*

     d. The **respiratory layer** contains the trachea, which bifurcates at the level of the sternal angle into the left and right main bronchi. The arch of the aorta passes over the left main bronchus, and the azygos vein passes over the right main bronchus. In **cross-sectional imaging**, the presence of the

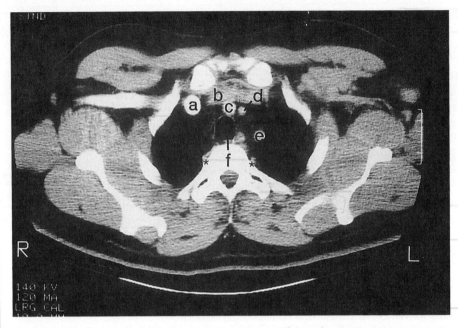

Figure 3–7a. Structures in the superior mediastinum (horizontal CT image at T_3). a. Right brachiocephalic vein, b. Left brachiocephalic vein, c. Brachiocephalic artery, d. Left common carotid artery, e. Left subclavian artery, f. Trachea. Asterisks: Sympathetic trunks.

trachea indicates that a cross-section can only be through the neck or the superior mediastinum and will not contain any part of the heart, which lies in the middle mediastinum inferior to the bifurcation of the trachea.

   e. The **digestive layer** contains the esophagus, which is indented on the left by the arch of the aorta and indented anteriorly by the left main bronchus.

     (1) These indentations are two sites where the esophagus is constricted.

     (2) Two other sites of constriction are at the origin of the esophagus in the neck and at the esophageal hiatus of the diaphragm.

CARCINOMAS OF THE ESOPHAGUS

Carcinomas of the esophagus commonly develop at one of the three sites of constriction in the mediastinum.

SWALLOWED FOREIGN BODY

A *swallowed foreign body* may become lodged at one of the three sites of constriction of the esophagus in the mediastinum.

   f. The **thoracic duct** courses posterior to the esophagus, then deviates to the left, and enters the root of the neck to drain lymph into the junction of the left subclavian and left internal jugular veins (Figure 3–2).

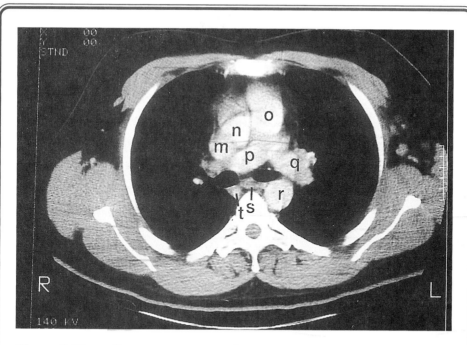

Figure 3–7b. m. Superior vena cava, n. Ascending aorta, o. Pulmonary trunk, p. Right pulmonary artery, q. Left pulmonary artery, r. Descending aorta, s. Esophagus, t. Azygous vein (horizontal CT at T_6).

g. The **left and right vagus nerves** pass through the superior mediastinum, enter the posterior mediastinum, and pass posterior to the root of the lung, where they form an esophageal plexus on the esophagus.
   (1) In the superior mediastinum, the left vagus nerve crosses the arch of the aorta and gives off the left recurrent laryngeal nerve, which hooks under the arch just posterior to the ligamentum arteriosum and courses superiorly to the larynx.
   (2) The right recurrent laryngeal nerve branches from the right vagus nerve in the root of the neck, hooks around the right subclavian artery, and is not in the superior mediastinum.
   (3) Pulmonary and cardiac branches of the vagus nerves enter the cardiac plexus.
h. The **left and right phrenic nerves** pass through the superior mediastinum, enter the middle mediastinum, pass anterior the root of the lung, and then descend to innervate the diaphragm.

ANEURYSM OF THE AORTIC ARCH

*An **aneurysm of the arch of the aorta** may compress the trachea, the esophagus, and the left recurrent laryngeal nerve. Patients may experience difficulty breathing, difficulty swallowing, and hoarseness.*

2. The **middle mediastinum** contains the pericardium, heart and adjacent great vessels, phrenic nerves, and accompanying pericardiacophrenic vessels.
   a. The **pericardium** consists of two layers: visceral and parietal (Figure 3–8).
      (1) The **visceral layer of serous pericardium** (epicardium) covers the heart.
      (2) The visceral layer is continuous with the parietal layer of serous pericardium at points where blood vessels enter or exit the heart.
   b. The **fibrous pericardium** consists of connective tissue, which forms a non-distensible external layer of parietal pericardium.
      (1) The fibrous pericardium is continuous with the adventitia of the vessels of the heart.
      (2) It is fused to the central tendon of the diaphragm.
   c. The **pericardial cavity** is a potential space between the parietal and visceral layers of serous pericardium. It contains a film of serous fluid, which permits the epicardium-covered heart to glide freely against the serous layer of parietal pericardium.
   d. The oblique and transverse pericardial sinuses are recesses of the pericardial cavity situated between the vessels of the heart.
      (1) The **transverse pericardial sinus** is found between the ascending aorta and the pulmonary trunk and superior vena cava.
      (2) The **oblique sinus** is found inferior to the pulmonary veins and to the left of the inferior vena cava.
   e. Sensory branches of the phrenic nerves innervate the parietal layers of pericardium.

CARDIAC TAMPONADE

Cardiac tamponade results from an *accumulation of fluid* in the pericardial cavity that compresses the chambers of the heart.

• *Pericardial effusion* may result in *Kussmaul's sign*, a distension of the veins of the neck on inspiration.

• *A penetrating wound of a chamber of the heart or weakening of a wall of the heart from a myocardial infarction may cause an acute tamponade as a result of accumulation of blood in the pericardial cavity (a **hemopericardium**).*

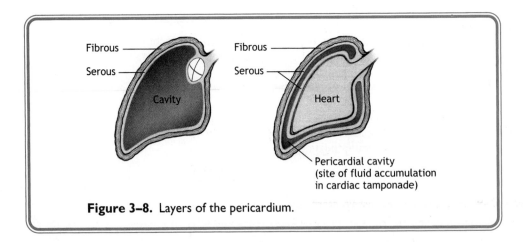

Figure 3–8. Layers of the pericardium.

- Patients with tamponade have decreased venous return and reduced cardiac output.
- In a **pericardiocentesis to relieve a tamponade**, a needle is passed through the parietal pericardium to aspirate blood from the pericardial cavity.
  – A site for a pericardiocentesis is at the left **xiphocostal angle.**
  – The needle enters the pericardial cavity after passing through skin, fascia, the rectus sheath, the rectus abdominis muscle, the fibrous layer, and the serous layer of parietal pericardium.

PERICARDITIS

Pericarditis causes a stiffening and reduced compliancy of the serous pericardium. In patients with pericarditis, the ventricles may not fill completely and cardiac output may be reduced because of a pericardial effusion.

3. The **heart** contains 4 chambers: 2 **atria** and 2 **ventricles**; each has a wall that consists of an outer epicardium, the myocardium and the fibrous skeleton, and an inner endocardium (Figure 3–9).

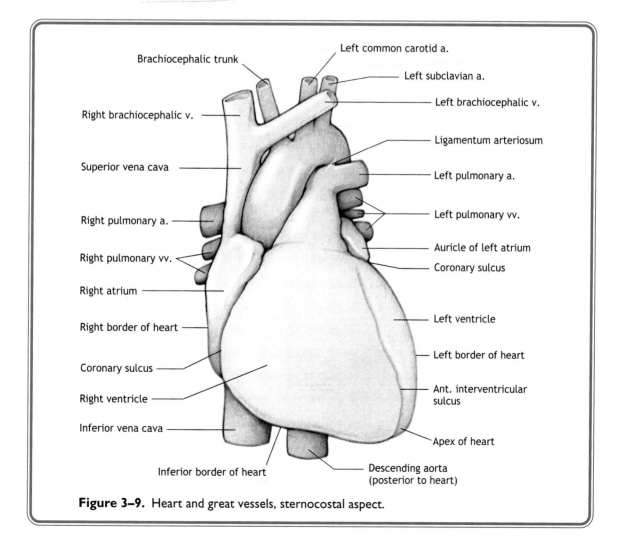

Figure 3–9. Heart and great vessels, sternocostal aspect.

a. The **epicardium** is the visceral layer of serous pericardium.
  b. The **myocardium** is formed by cardiac muscle, which is most prominent in the ventricles.
  c. The **fibrous skeleton** provides an attachment for the myocardium, maintains the patency of the atrioventricular and semilunar valves, provides an attachment site for the leaflets and cusps of the valves, and separates the atria and the ventricles to allow the atria to contract independently of the ventricles.
  d. The **endocardium** lines the chambers of the heart.
4. The **heart has 6 surfaces**, including a base and an apex.
  a. The **base** forms the posterior surface and is formed mainly by the left atrium and the 4 pulmonary veins entering the left atrium. The left atrium is separated by the pericardium, the descending aorta, and the esophagus from the bodies of the T6 through T8 vertebrae.
  b. The **apex** is formed by the tip of the left ventricle that is found posterior to the left fifth intercostal space.
  c. The **right surface** of the heart is formed by the right atrium, which is situated between the superior and the inferior vena cavae.

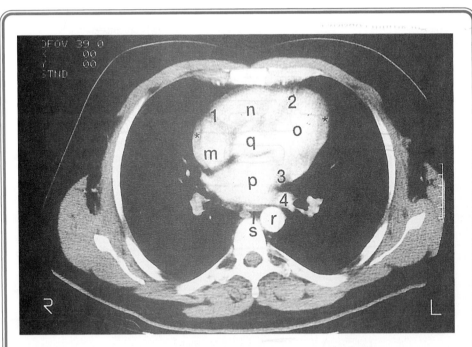

Figure 3–10a. Structures in the middle mediastinum (horizontal CT image at T_8): m. Right atrium, n. Right ventricle, o. Left ventricle, p. Left atrium, q. Ascending aorta, r. Descending aorta, s. Esophagus. 1. Location of right coronary artery, 2. Location of anterior interventricular artery, 3. Location of circumflex branch of left coronary artery, 4. Left pulmonary vein. Asterisks: Approximate locations of the phrenic nerves.

d. The **left surface** of the heart is formed by the left ventricle and the auricle of the left atrium.

e. The **anterior or sternocostal surface** is formed by the right ventricle; part of the right ventricle extends to the left of the sternum.

f. The **inferior or diaphragmatic surface** is formed by the left and the right ventricles.

5. The **heart has four borders** that correspond to the margins of the heart and its blood vessels as viewed in a posteroanterior chest radiograph (Figures 3–9, 3–10a, b).

a. The **right border** is formed by the right atrium and the superior and inferior vena cavae.

b. The **left border** is formed by the left ventricle and the left auricle.

c. The **inferior border** is formed by the juncture between the anterior surface and the diaphragmatic surface of the heart and is formed mainly by the right ventricle.

d. The **superior border** is formed by the left and right atria and by the ascending aorta, the pulmonary trunk, and the superior vena cava.

6. The individual **chambers of the heart** are the right atrium, right ventricle, left atrium, and left ventricle (Figures 3–10a, b, 3–11).

a. The **right atrium**

(1) The right atrium consists of a smooth-walled part (the **sinus venarum**) and a rough-walled part (including the anterior wall and the auricle)

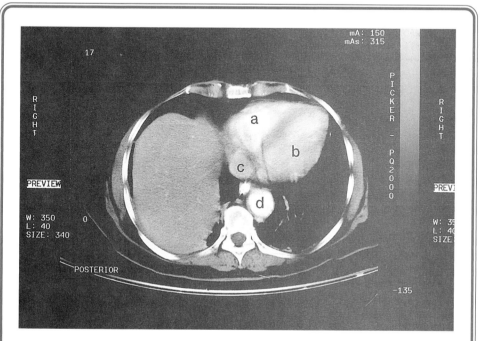

Figure 3–10b. a. Right ventricle, b. Left ventricle, c. Inferior vena cava, d. Descending aorta.

that contains **pectinate** muscles. The **crista terminalis** separates the sinus venarum from the part of the atrium containing pectinate muscles.

(2) It contains the **fossa ovalis** in the interatrial septum. The fossa ovalis is a remnant of the foramen ovale, which during fetal life permits oxygenated blood to bypass the lungs by shunting blood from the right atrium to the left atrium.

ATRIAL SEPTAL DEFECTS

*A **small patency (a probe patency)** in the upper part of the fossa ovalis of the interatrial septum may be present but is not symptomatic. A **large patency** in the fossa ovalis may form a symptomatic atrial septal defect.*

- *Normally, blood pressure is higher in the left side of the heart than in the right side of the heart in postnatal life.*
- *In patients with an atrial septal defect, blood from the left atrium will be shunted through the defect into the right atrium.*

    (3) The right atrium contains the openings of the superior vena cava, the inferior vena cava, the ostium of coronary sinus, and the openings of anterior cardiac veins. All of these structures open into the sinus venarum and deliver poorly oxygenated venous blood to the right atrium from the systemic and cardiac circulations.

    (4) It also contains the right atrioventricular orifice, which opens into the right ventricle and is closed by the three leaflets of the **tricuspid valve** (Figure 3–11).

  b. The **right ventricle**

    (1) The right ventricle consists of a rough-walled part with thick muscular walls that contain trabeculae carneae.

    (2) It consists of a smooth-walled part, the **conus arteriosus** or infundibulum. The conus arteriosus leads to the pulmonary trunk; the

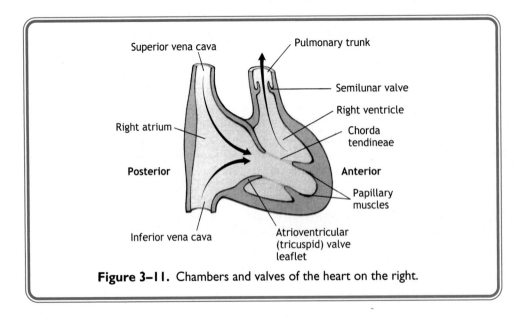

Figure 3–11. Chambers and valves of the heart on the right.

pulmonic semilunar valve is at the junction between the infundibulum and the pulmonary trunk.

(3) The right ventricle contains three **papillary muscles** (anterior, posterior, and septal) and **chordae tendineae** that attach to the papillary muscles and to the edges of the anterior, posterior, and septal leaflets of the tricuspid valve.

(4) It is separated from the left ventricle by an **interventricular septum**, which has a membranous component and a muscular component.

(5) The right ventricle contains the **septomarginal trabecula or moderator band**, which extends from the base of the anterior papillary muscle to the interventricular septum. The septomarginal trabecula contains the right bundle branch of the interventricular bundle of His, a part of the cardiac conduction system.

VENTRICULAR SEPTAL DEFECT

*A large **postnatal defect** in the interventricular septum (most often in the membranous part) results in too much pulmonary blood flow caused by a shunt of blood from the left ventricle into the right ventricle. In patients with a ventricular defect, pulmonary hypertension may result, causing congestive heart failure.*

    c. The **left atrium**

      (1) The left atrium consists of a smooth-walled part and a rough-walled part, the auricle, which contains pectinate muscles.

      (2) It contains the openings of **4 pulmonary veins** on the posterior wall.

      (3) The left atrium contains the left atrioventricular orifice, which opens into the left ventricle, and is closed by the 2 leaflets of the **mitral valve**.

    d. The **left ventricle**

      (1) The left ventricle consists of a muscular wall with **trabeculae carneae** that is twice as thick as the wall of the right ventricle.

      (2) It contains a smooth-walled part, the **aortic vestibule**, which leads to the ascending aorta. The **aortic semilunar valve** is at the junction between the vestibule and the ascending aorta. The valve contains 3 cusps—a right, a left, and a posterior, an aortic sinus distal to each cusp.

      (3) The left ventricle has 2 **papillary muscles** (anterior and posterior) with chordae tendineae that attach to the edges of the anterior and posterior leaflets of the mitral valve. Papillary muscles in both ventricles contract during ventricular contraction to keep the tricuspid and mitral valves closed.

7. **Heart valve sounds** are normal sounds caused by the acceleration and deceleration of blood flowing through a valve, and are auscultated adjacent to the sternum just downstream from the valve.

    a. The first heart sound (S1) corresponds to the closing of the mitral and tricuspid valves, and occurs approximately at the beginning of systole.

      (1) The **tricuspid valve** is heard adjacent to the left side of the sternum in the fifth intercostal interspace.

      (2) The **mitral valve** is heard in the fifth intercostal interspace in the midclavicular line (the apex beat).

b. The second heart sound (S2) corresponds to the closing of the aortic and pulmonic valves and occurs approximately at the end of systole and at the beginning of diastole.

      (1) The aortic and pulmonic valves are closed by the backflow of blood, preventing blood from flowing back into a ventricle.

      (2) The **aortic valve** is heard adjacent to the right side of the sternum in the second intercostal interspace.

      (3) The **pulmonic valve** is heard adjacent to the left side of the sternum in the second intercostal interspace.

HEART MURMURS AND VALVULAR HEART DISEASE

A *heart murmur* is an abnormal sound that results from vibrations produced by the **turbulent flow of blood**.

- *Murmurs in valvular heart disease result when a valve is not fully closed (an insufficient or incompetent valve) or from a decrease in valve diameter (a stenotic valve).*

- *For most of ventricular systole the mitral valve should be closed and the aortic valve should be open, so that "systolic valvular defects" include mitral insufficiency and aortic stenosis. For most of ventricular diastole the mitral valve should be open and the aortic valve should be closed, so that "diastolic valvular defects" include mitral stenosis and aortic insufficiency.*

    8. Coronary arteries and their branches supply arterial blood to the heart; **cardiac veins** drain the heart (Figures 3–10, 3–12).

      a. The right and left coronary arteries arise from the right and left aortic sinuses of the ascending aorta, which are located just distal to the right and left semilunar cusps of the aortic valve.

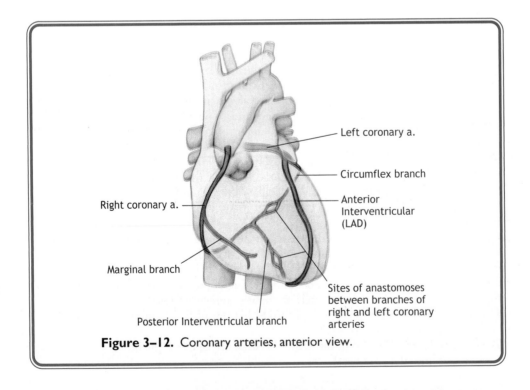

Figure 3–12. Coronary arteries, anterior view.

b. The right coronary artery
   (1) The **right coronary artery** is a long artery that courses in the coronary sulcus and supplies the right atrium and the right ventricle.
   (2) It typically has **nodal branches** that supply the sinoatrial and atrioventricular nodes.
   (3) The right coronary has a **marginal branch** that supplies the right ventricle.
   (4) It is dominant if it gives rise to the **posterior interventricular artery** (the typical pattern). The posterior interventricular artery courses in the posterior interventricular sulcus and supplies the right and left ventricles and the posterior part of the interventricular septum.
c. The left coronary artery
   (1) The **left coronary artery** is a short artery that branches within 1 cm of its origin from the left aortic sinus into the anterior interventricular artery and the circumflex artery.
   (2) The **anterior interventricular (or left anterior descending) artery** courses in the anterior interventricular sulcus and supplies the right and left ventricles and two-thirds of the interventricular septum, including the atrioventricular bundle.
   (3) The **circumflex artery** courses in the coronary sulcus and supplies the left atrium and the left ventricle.
   (4) A **marginal branch** arises from the circumflex artery and supplies the left ventricle.
d. The **coronary arteries** are functional end arteries that establish **collateral circulation** with each other through anastomoses:
   (1) Between the anterior and posterior interventricular arteries in the posterior part of the interventricular sulcus in a right dominant heart.
   (2) Between the circumflex branch of the left coronary artery and the right coronary artery in the posterior part of the coronary sulcus.

ANGINA PECTORIS

Angina pectoris is **chest pain** that results from transient ischemia brought on by exertion.

- The ischemia that causes angina pectoris results from reduced blood flow to cardiac muscle because of a narrowing of a coronary artery, but there is no loss of cardiac muscle cells. Patients have substernal pain that may be referred over the T1-5 dermatomes of the thoracic wall that correspond to the same segments of the spinal cord that provide the sympathetic innervation to the heart.
- Referred pain may be felt in the T1 dermatome in the medial aspect of the left arm and forearm and may be felt over cervical dermatomes in the neck, up to the level of the angle of the mandible.

MYOCARDIAL INFARCTION

A **myocardial infarction** results from a localized **avascular necrosis of cardiac muscle cells** caused by prolonged ischemia.

- The anterior interventricular artery is a common site of an occlusion that results in an acute myocardial infarction. Less frequently occluded are the right coronary artery and the circumflex branch of the left coronary artery.
- In patients, the onset of a myocardial infarction is usually marked by sudden, severe pain beneath the sternum.

e. **Cardiac veins** course with branches of the coronary arteries and drain mainly into the coronary sinus, the main vein of the heart.
   (1) The **great cardiac vein** courses with the anterior interventricular artery and drains into the coronary sinus.
   (2) The **middle cardiac vein** courses with the posterior interventricular artery and drains into the coronary sinus.
   (3) The **small cardiac vein** courses with the marginal branch of the right coronary artery and drains into the coronary sinus.
   (4) The **anterior cardiac veins** are exceptions; they drain the right ventricle and empty directly into the right atrium.
9. **The cardiac conduction system** (Figure 3–13).
   a. The cardiac conduction system initiates and spreads a wave of depolarization across the cardiac muscle in both atria and both ventricles.
   b. It consists of modified cardiac muscle cells in the sinoatrial (SA) node and in the atrioventricular (AV) node, the atrioventricular bundle of His, and Purkinje fibers.
   c. The **SA node**
      (1) The SA node is located in the sinus venarum of the right atrium near the entrance of the superior vena cava.
      (2) It acts as the **pacemaker** because it contains cells that depolarize at a faster rate than other cardiac muscle cells. The SA node depolarizes approximately 70 times/min at rest.

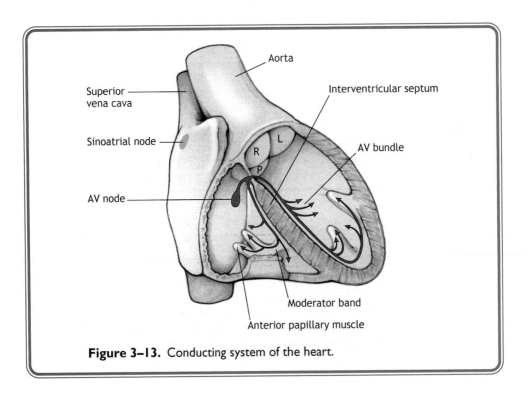

Figure 3–13. Conducting system of the heart.

(3) The SA node initiates a wave of depolarization that spreads to cardiac muscle cells in both atria, causing the atria to contract.

d. The **AV node**

(1) The AV node is located in the interatrial septum near the right atrioventricular orifice.

(2) It is stimulated by depolarization of cardiac muscle cells in the atria and **delays the wave of depolarization** from passing to the ventricles to allow both atria to fully contract. The AV node depolarizes approximately 50–60 times/min at rest.

e. The **AV bundle of His**

(1) The AV bundle of His arises from cells of the AV node and passes into the interventricular septum.

(2) It divides into right and left bundle branches that course through the interventricular septum; **Purkinje fibers** spread a wave of depolarization to cardiac muscle cells in the right and left ventricles.

(3) The AV bundle of His causes each of the papillary muscles in the ventricles to contract to keep the AV valves shut when the ventricles contract. The **septomarginal trabecula or moderator band** carries Purkinje fibers from the right bundle branch to the base of the anterior papillary muscle.

[handwritten margin note: In charge of papillary muscle contraction so valve closing]

AV BLOCK

*In an **atrioventricular** or **heart block**, conduction is slowed through the AV node, some impulses are not transmitted through the node, or, in a complete block, no impulses are conducted through the AV node.*

- *In a **complete AV block**, the contractions of the atria and the ventricles become dissociated, and these chambers beat independently.*

- *The atria may continue to contract about 70 times/min; a pacemaker may develop in the AV bundle distal to the site of the block, initiating contraction of the ventricles at a rate of 30–40 times/min.*

10. Innervation of the heart

a. **Sympathetic and parasympathetic nerves** innervate the conduction system and the cardiac muscle of the heart.

b. Both types of autonomic fiber reach the heart by passing through either the superficial cardiac plexus situated inferior to the arch of the aorta or the deep cardiac plexus situated anterior to the bifurcation of the trachea.

c. Each **cardiac plexus contains autonomic fibers**, including

(1) Preganglionic parasympathetic axons from the vagus nerves (CN X) that synapse in terminal ganglia.

(2) Postganglionic parasympathetic axons from terminal ganglia in the plexus; these fibers synapse in the SA and AV nodes and in the walls of the coronary arteries.

(3) Postganglionic sympathetic axons from cervical and upper thoracic splanchnic nerves, which have their cell bodies in the superior, middle, and inferior cervical paravertebral ganglia and in the T1 through T5 paravertebral ganglia; postganglionic sympathetic axons synapse in the SA and AV nodes and in the walls of the coronary arteries.

  d. Each **cardiac plexus contains sensory fibers.**
   (1) Sensory fibers **innervate baroreceptors and chemoreceptors** in the heart and in the arch of the aorta that are sensitive to changes in blood pressure or to the chemical content of the blood, respectively. These fibers enter the CNS after coursing with the vagus nerves.
   (2) They also **carry pain sensations** back into the CNS with sympathetic nerves and enter the CNS from the T1 through T5 spinal cord segments. Pain associated with angina pectoris may be referred over the T1 through T5 dermatomes.
  e. The **sympathetic innervation accelerates the heart rate** and increases the force of cardiac contraction mainly by increasing the rate of depolarization of cells in the SA and AV nodes. It also inhibits smooth muscle in the coronary arteries, which allows the arteries to dilate.
  f. The **parasympathetic innervation slows heart rate**, decreases the force of cardiac contraction, and promotes vasoconstriction of the coronary arteries.

VIII. Posterior Mediastinum

 A. The **posterior mediastinum** is inferior to the horizontal plane that intersects the sternal angle and the disk between the T4 and T5 vertebrae.
 B. It is found between the pericardium, the vertebral column, and the diaphragm.
  1. The posterior mediastinum contains structures that are also found in other divisions of the mediastinum.
   a. The **descending aorta** begins at the horizontal plane, where it is continuous with the arch of the aorta in the superior mediastinum. It gives rise to posterior intercostal arteries that supply the thoracic wall, bronchial arteries that supply lung tissue, and esophageal arteries.
   b. The **thoracic duct** arises from the cisterna chyli below the diaphragm and courses through the posterior and the superior mediastinum to terminate in the root of the neck on the left at the origin of the left brachiocephalic vein.
   c. The **esophagus** courses through the superior and posterior mediastina. It contains skeletal muscle in its upper third, smooth muscle in its lower third, and a mix of both types of muscle in its middle third. The esophagus is innervated by the esophageal plexus, which is formed mainly by the right and left vagus nerves. Just superior to the diaphragm the plexus forms an anterior vagal trunk from fibers of the left vagus nerve and a posterior vagal trunk from fibers of the right vagus nerve.
   d. The azygos and 1 or 2 hemiazygos veins (Figure 3–3) receive posterior intercostal veins from the right and left side of the thoracic wall, respectively. The hemiazygos veins drain into the azygos vein after crossing the midline at about the level of the T8 vertebra. The azygos vein drains into the superior vena cava.

IX. Anterior Mediastinum

 A. The **anterior mediastinum** is found below the transverse plane between the body of the sternum and the pericardium.
 B. The main structure in the anterior mediastinum is the **thymus**.
  1. Thymomas are common mediastinal tumors.
  2. A thymoma may be found in either the anterior or the superior mediastinum.

CLINICAL PROBLEMS

A thin 26-year-old man is an avid skin diver. During a diving trip in the ocean, he surfaces and complains of difficulty breathing and chest pain. There are no indications of a surface wound of the chest. In the emergency room, imaging reveals a collapsed lung on the right. Cardiac output is normal.

1. These signs and symptoms indicate that patient may have:

   A. An open pneumothorax

   B. A tension pneumothorax

   C. A spontaneous pneumothorax

   D. Cardiac tamponade

   E. A Pancoast tumor

The pain experienced by the patient in Question 1 may have resulted from irritation of pain fibers that innervate the costal pleura.

2. What nerves carry pain from the costal pleura?

   A. Vagus nerves

   B. Greater splanchnic nerves

   C. Intercostal nerves

   D. Phrenic nerves

   E. Cardiopulmonary splanchnic nerves

An 18-year-old man is diagnosed to have high blood pressure during a physical. The blood pressure is significantly higher in both upper limbs than in both lower limbs. Imaging reveals bilateral erosion of the anterior and lateral parts of his ribs. Angiography reveals a narrowing of the aorta.

3. Where is the most likely site of the aortic constriction?

   A. Between the brachiocephalic trunk and the left common carotid artery

   B. Just distal to the ligamentum arteriosum

   C. Between the origin of the subclavian artery and the ligamentum arteriosum

   D. In the middle mediastinum

   E. At the point where the arch of the aortic indents the esophagus

4. In the patient in Question 3, in which of the following vessels is there a retrograde flow of blood?

   A. Anterior intercostal arteries

   B. Posterior intercostal arteries

   C. Internal thoracic arteries

   D. Right subclavian artery

   E. Descending aorta

A 55-year-old male lawyer is brought to the emergency room by his wife. The wife states that her husband complained of a sharp, squeezing chest pain behind the sternum after a meal and has had repeated episodes of chest pain after exertion over the past several years. A diagnosis is made of an acute myocardial infarction (MI) of the interventricular bundle.

5. What was the most likely site of an occlusion?

 A. Posterior interventricular artery

 B. Circumflex artery

 C. Marginal artery

 D. Right coronary artery

 E. Anterior interventricular artery

6. What part of the cardiac conduction system might have been affected in the patient in Question 5 if the right coronary artery is the site of an occlusion?

 A. Stellate ganglion

 B. Cardiac plexus

 C. Sinoatrial node

 D. Atrioventricular bundle

 E. Left vagus nerve

7. Which of the following is a neural structure that carries visceral pain fibers from the heart that results in referred pain over the T1-5 dermatomes?

 A. Ventral roots of the T1-5 spinal nerves

 B. Dorsal roots of the T1-5 spinal nerves

 C. Greater splanchnic nerves

 D. Gray rami communicantes of the T1-5 spinal nerves

 E. Dorsal rami of the T1-5 spinal nerves

A 64-year-old woman is admitted to the hospital complaining of shortness of breath and difficulty swallowing. The patient coughs up blood during the examination and speaks with a hoarse voice. Diagnosis is made of an esophageal carcinoma that commonly occurs at a site of constriction of the esophagus where it is indented by the left main bronchus.

8. What other anatomic structure indents the esophagus and causes a constriction?

 A. Arch of the aorta

 B. Inferior vena cava

 C. Azygos vein

 D. Superior vena cava

 E. Left ventricle

9. In Question 8, the patient's hoarse voice may have been caused by the mass imping-ing on:

 A. The left phrenic nerve

 B. The left recurrent laryngeal nerve

C. The left vagus nerve

D. The sympathetic trunk on the left

E. Cardiopulmonary splanchnic nerves

10. If the esophageal mass expands anteriorly into the middle mediastinum, what structure might initially become compressed?

   A. Left ventricle

   B. Right ventricle

   C. Right atrium

   D. Left atrium

A patient has been admitted to the hospital after suffering from a knife wound of the chest just to the left of the sternum. He is slightly cyanotic, and there is a distension of veins of the neck during inspiration. You suspect that the patient has a cardiac tamponade and order a pericardiocentesis.

11. What is the last tissue layer that the needle must traverse in order to reach the accumulating blood?

   A. Epicardium

   B. Fibrous pericardium

   C. Mediastinal pleura

   D. Visceral pericardium

   E. Serous layer of parietal pericardium

12. A surgical procedure is performed in a patient to remove a thymic tumor. Which of the following structures must be avoided during the surgery that is directly posterior to the thymoma in the superior mediastinum?

   A. Arch of the aorta

   B. Esophagus

   C. Trachea

   D. Left brachiocephalic vein

   E. Left common carotid artery

A thoracic CT scan of your patient reveals that a neoplasm of the right lung has expanded medially into the wall of the pericardium, and has compressed nerve fibers that course on the surface of the pericardium anterior to the root of the lung.

13. Which of the following signs or symptoms might you expect to see in this individual?

   A. Decreased heart rate and cardiac output

   B. Hoarseness

   C. Horner's syndrome

   D. Slow gastric emptying

   E. Ascension of the right side of the diaphragm during inspiration

In the OR, a cardiac surgeon is performing a median sternotomy to gain access to the heart for a coronary artery bypass. While cutting through the sternum, a large quantity of blood suddenly erupts from the chest. The cardiac surgeon immediately suspects that the saw has lacerated a structure in the mediastinum posterior to the body of the sternum.

14. What structure is most likely to be injured?

   A. The azygos vein

   B. The arch of the aorta

   C. The left atrium

   D. The right atrium

   E. The right ventricle

ANSWERS

1. The answer is C. Because there was no evidence of a surface wound, an emphysematous bleb spontaneously may have ruptured, and air was introduced into the pleural sac through the visceral pleura.

2. The answer is C. The costal parietal pleura is innervated by intercostal nerves.

3. The answer is B. The most common site of a coarctation of the aorta in adults is just distal to the ligamentum arteriosum.

4. The answer is B. Anastomoses between the anterior intercostal arteries and posterior intercostal arteries provide collateral circulation that bypasses the coarctation. Blood flows in the retrograde direction through the posterior intercostal arteries into the descending aorta.

5. The answer is E. The anterior interventricular artery supplies most of the interventricular septum.

6. The answer is C. The SA node is supplied by the right coronary artery.

7. The answer is B. Dorsal roots of the T1-5 spinal nerves carry visceral pain fibers into the spinal cord. Visceral pain fibers from the heart would not course in any of the other listed structures (see Figure 1–3).

8. The answer is A. The esophagus is indented on the left by the arch of the aorta in the superior mediastinum.

9. The answer is B. The left recurrent laryngeal nerve hooks under the arch of the aorta in the mediastinum and courses in close proximity to the esophagus.

10. The answer is D. The left atrium is on the posterior surface of the heart and directly anterior to the esophagus.

11. The answer is E. The needle enters the pericardial sac after passing through the serous layer of parietal pericardium.

12. The correct answer is C. The left brachiocephalic vein is directly posterior to the thymus in the superior mediastinum.

13. The correct answer is D. The right phrenic nerve courses on the pericardium anterior to the root of the lung. Phrenic nerve lesions result in a paradoxical elevation of one side of the diaphragm during inspiration.

14. The answer is E. The right ventricle lies directly posterior to the body of the sternum.

CHAPTER 4
ABDOMEN

I. The **anterolateral abdominal** wall can be divided into **9 regions or 4 quadrants** useful for palpating abdominopelvic organs and localizing pain associated with abdominopelvic disease (Figure 4–1).

   A. The **9 regions** of the anterolateral abdominal wall are bounded by 4 intersecting planes, 2 vertical midclavicular planes, a horizontal subcostal plane, and a horizontal transtubercular plane.

      1. The **midclavicular planes** pass through the midpoint of the clavicle and the midpoint of the inguinal ligament. They separate the epigastric, umbilical, and hypogastric regions from the hypochondriac, lumbar, and inguinal regions.

      2. The **subcostal plane** passes just inferior to the costal margin and intersects the L3 vertebra.

         a. The L3 vertebra marks the inferior extent of the third part of the duodenum and the origin of the inferior mesenteric artery from the abdominal aorta.

         b. The epigastric region and the umbilical region are found above and below the subcostal plane, respectively. The umbilical region contains the T10 dermatome, a common site of referred pain.

      3. The **transtubercular plane** intersects the iliac tubercles and the body of the L5 vertebra. The umbilical region and the hypogastric region are found above and below the transtubercular plane, respectively.

   B. The **4 quadrants** are bounded by the median plane and by a transumbilical plane that intersects the disk between the L3 and the L4 vertebrae.

      1. The right and left upper quadrants are superior to the transumbilical plane.

      2. The right and left lower quadrants are inferior to the transumbilical plane.

   C. The **transpyloric plane** that intersects the L1 vertebra is an important level for imaging. The transpyloric plane contains the pyloric part of the stomach (variable), the first part of the duodenum, the fundus of the gallbladder, the neck of the pancreas, the origin of the superior mesenteric artery, the hepatic portal vein, and the splenic vein.

II. The **abdominal wall consists of 8 tissue layers;** the lumbar vertebrae, rib cage, and bony pelvis also contribute to the wall and provide attachment sites for muscles of the wall.

   A. The 8 tissue layers of the anterior abdominal wall, from superficial to deep, include skin, several layers of fascia, 3 layers of muscle, and the peritoneum (Figure 4–2).

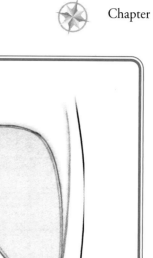

Figure 4–1. The 9 regions of the abdomen.

1. The **superficial fascia**, which is deep to the skin, consists of an outer layer of **Camper's fascia** that contains fat and is continuous with superficial fascia everywhere and an inner layer of **Scarpa's fascia** that lacks fat.
   a. Scarpa's fascia is continuous with superficial perineal fascia (Colles' fascia) covering the perineum.
   b. It is also continuous with superficial penile fascia covering the penis and clitoris.
   c. Scarpa's fascia is continuous with **dartos fascia** of the scrotum, which contains the **dartos muscle**, a smooth muscle that functions to help regulate the thermal environment of the testis.
2. The **superficial epigastric veins** of the superficial fascia drain into the femoral vein and communicate with small paraumbilical veins at the umbilicus.

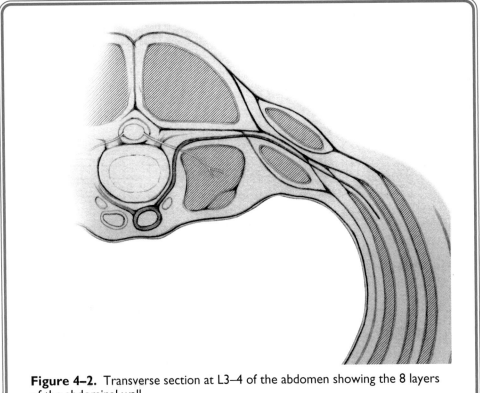

Figure 4–2. Transverse section at L3–4 of the abdomen showing the 8 layers of the abdominal wall.

VARICOSE SUPERFICIAL EPIGASTRIC VEINS

Obstruction of either the inferior vena cava or the hepatic portal vein, both of which drain structures below the diaphragm, may result in varicosities of the superficial epigastric veins.

CLINICAL
CORRELATION

B. The **external oblique muscle** consists of fibers that run medially and inferiorly.
   1. This muscle becomes the external oblique aponeurosis at the midclavicular line.
   2. The aponeurosis attaches inferiorly and medially to the pubic tubercle and inferiorly and laterally to the anterior superior iliac spine (ASIS).
      a. The aponeurosis is folded posteriorly between the pubic tubercle and the ASIS and forms the **inguinal ligament**. The inguinal ligament marks the inferior border of the anterior abdominal wall.
      b. The aponeuroses of the external oblique muscles contribute to the **rectus sheath** and interdigitate in the anterior midline at the linea alba.
C. The **internal oblique muscle** attaches to the lateral two-thirds of the inguinal ligament; its fibers run superiorly and medially, perpendicular to those of the external oblique.
   1. This muscle becomes the internal oblique aponeurosis at the midclavicular line. The aponeuroses of the internal oblique muscles contribute to the rectus sheath and interdigitate in the midline at the linea alba.

2. The internal oblique muscle gives rise to the **cremaster muscle**, which covers the testis and the spermatic cord.

D. The **transversus abdominis muscle** consists of fibers that run horizontally. The aponeuroses of the transversus abdominis muscles contribute to the rectus sheath and interdigitate in the midline at the linea alba.

E. **Transversalis fascia** is deep to the transversus abdominis muscle and in contact with the rectus abdominis muscle below the **arcuate line**. It gives rise to **internal spermatic fascia**, which covers the testis and the spermatic cord.

F. **Extraperitoneal fat** is the tissue layer in which the ovaries and testes develop.

G. **Peritoneum** is the serous membrane that lines the walls of the abdominopelvic cavity and invests gastrointestinal structures below the diaphragm.

   1. The peritoneum consists of 2 layers.
     a. A **parietal layer** lines the abdominopelvic walls.
     b. A **visceral layer** covers gastrointestinal structures.

   2. The **peritoneal cavity** is a potential space between the parietal and visceral layers of peritoneum.
     a. In males the peritoneal cavity is completely closed.
     b. In females the uterine tubes open into the peritoneal cavity.

ASCITES

Ascites is an accumulation of fluid in the peritoneal cavity and may be caused by peritonitis or result from congestion of the venous drainage of the abdomen.

   3. The peritoneum forms folds or reflections known as **mesenteries**; a mesentery consists of 2 layers of peritoneum.
     a. Each layer of a mesentery is continuous with **visceral peritoneum**, which covers a gastrointestinal structure, and with parietal peritoneum, which lines the abdominal wall.
     b. The **dorsal mesentery** reflects off of the midline of the dorsal body wall; during fetal life, the dorsal mesentery suspends gastrointestinal structures below the diaphragm.
     c. The **ventral mesentery** reflects off of the midline of the ventral body wall; during fetal life, the ventral mesentery suspends gastrointestinal structures derived from the fetal foregut and the spleen.
     d. A mesentery contains blood vessels, lymphatic vessels, and nerves that supply structures suspended by the mesentery.

   4. Parietal peritoneum forms 3 folds on the anterior abdominal wall that contain remnants of fetal structures and 2 folds that contain patent structures in postnatal life (Figure 4–3).
     a. The **median umbilical fold** contains the urachus, a remnant of the allantois, which connected the fetal bladder to the umbilicus.
     b. **Two medial umbilical folds** contain medial umbilical ligaments, the obliterated remnants of the fetal umbilical arteries that returned fetal blood to the placenta.
     c. **Two lateral umbilical folds** contain the inferior epigastric blood vessels; these vessels enter the rectus sheath and supply the rectus abdominis muscle.

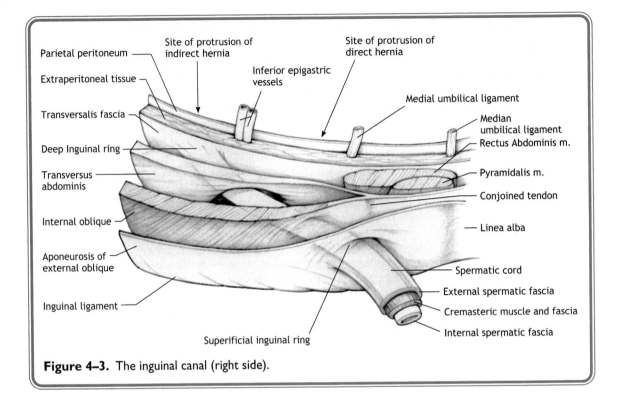

Figure 4–3. The inguinal canal (right side).

    H. Lower intercostal nerves (T7 through T11), the **subcostal nerve** (T12), and the **iliohypogastric and ilioinguinal nerves** (L1) innervate skin, fascia, and muscles of the anterolateral abdominal wall.

III. The **rectus sheath** is formed by the aponeuroses of the 3 muscles of the anterolateral abdominal wall and contains the rectus abdominis muscle.

    A. The **external oblique aponeurosis** forms the anterior wall of the sheath. In the superior three quarters of the sheath, the **internal oblique aponeurosis** splits to form part of both the anterior and the posterior walls, and the **transversus abdominis aponeurosis** forms part of the posterior wall.

    B. At the **arcuate line**, the aponeuroses of the internal oblique and the transversus abdominis muscles pass anterior to the rectus abdominis muscle, leaving the posterior wall of the rectus in contact with transversalis fascia.

        1. The **arcuate line** is located between the umbilicus and the pubic symphysis.

        2. The **inferior epigastric artery**, a branch of the external iliac artery, and the superior epigastric artery, a branch of the internal thoracic artery, supply the rectus abdominis muscle.

    C. The **rectus abdominis muscle** flexes the lumbar vertebrae, and, with the other muscles of the abdominal wall, compresses the abdomen, thereby increasing intra-abdominal pressure.

IV. The **inguinal region** of the anterior abdominal wall contains the inguinal canal (see Figure 4–3).

A. The **inguinal canal** is formed by 4 of the 8 tissue layers of the anterior abdominal wall; outpocketings of 3 of these layers give rise to **spermatic fasciae**, which cover the testis and structures in the **spermatic cord**.

B. The inguinal canal has a posterior wall, an anterior wall, a floor, and a roof.

1. The **posterior wall** of the inguinal canal

    a. The posterior wall of the inguinal canal is formed by **transversalis fascia** and is the first tissue layer that covers the testis as it enters the inguinal canal.

    b. It contains the **deep inguinal ring**, a fault in the transversalis fascia, which forms the posterior opening of the inguinal canal; the inferior epigastric artery and vein are medial to the deep inguinal ring.

    c. It also gives rise to **internal spermatic fascia**, an outpocketing of transversalis fascia at the deep inguinal ring.

2. The **anterior wall** of the inguinal canal

    a. The anterior wall of the inguinal canal is formed by the **aponeurosis of the external oblique muscle** and is the traversed by the testis as it exits the inguinal canal.

    b. It contains the **superficial inguinal ring**, a fault in the external oblique aponeurosis that forms the anterior opening of the inguinal canal; the **inferior epigastric artery and vein** are lateral to the superficial inguinal ring.

    c. It also gives rise to **external spermatic fascia**, an outpocketing of the external oblique fascia at the superficial inguinal ring.

3. The **floor of the inguinal canal** is formed by the inguinal ligament.

4. The **roof of the inguinal canal** is formed by the arched muscle fibers of the **transversus abdominis** and **internal oblique muscles.** The aponeuroses of these muscles fuse to form the **conjoined tendon** that attaches to the pubic bone posterior to the superficial inguinal ring.

5. Cremasteric fascia is derived from the fascia of the internal oblique, and the cremaster muscle consists of fibers of the internal oblique muscle that extend along the spermatic cord. Contraction of the **cremaster muscle** elevates the testis and helps regulate the thermal environment of the testis.

6. The testis passes under the transversus abdominis muscle; thus, the transversus does not contribute to the spermatic fasciae.

CREMASTERIC REFLEX

The **cremasteric reflex** utilizes sensory and motor fibers in the ventral ramus of the L1 spinal nerve.
- Stroking the skin of the superior and medial thigh stimulates sensory fibers of the **ilioinguinal nerve**.
- **Motor fibers** from the genital branch of the **genitofemoral nerve** cause the cremaster muscle to contract, elevating the testis.

TORSION

Torsion of a testicle results in a sudden onset of testicular pain and a loss of the cremasteric reflex.

C. The **inguinal canal** in males contains the spermatic cord, which contains a number of structures.

1. The **ductus deferens** conveys sperm from the epididymis to the ejaculatory duct in the male pelvis.
2. The **testicular artery** arises from the abdominal aorta between the L2 and the L3 vertebrae, and the **artery to the ductus deferens** arises from a branch of the internal iliac artery.
3. The spermatic cord also contains the **pampiniform plexus of the testicular vein**.
   a. The right testicular vein drains into the inferior vena cava, and the left testicular vein drains into the left renal vein.
   b. The pampiniform plexus, the cremaster muscle, and the dartos muscle provide thermoregulatory control for the testis so that spermatogenesis occurs at a constant temperature.
4. **Sympathetic nerves** from the T11 to L1 spinal cord segments innervate smooth muscle in the **ductus deferens**.
   a. Smooth muscle in the **ductus deferens** contracts during emission.
   b. This contraction is part of a **sexual reflex** that delivers sperm and seminal fluid to the prostatic urethra.
5. Another nerve in the spermatic cord is the **genital branch of the genitofemoral nerve**, which innervates the cremaster muscle.
6. **Lymphatic vessels in the spermatic cord** drain the testis. Testicular lymphatic vessels pass through the inguinal canal and drain directly to **lumbar nodes** in the posterior abdominal wall (see Figure 1–6).
7. The spermatic cord contains a remnant of the **processus vaginalis**.
   a. The **processus vaginalis** is an evagination of parietal peritoneum that descends through the inguinal canal during fetal life.
   b. The **tunica vaginalis** is a patent remnant of the processus vaginalis that covers the anterior and lateral parts of the testis.

ABNORMAL CYSTS IN THE SPERMATIC CORD

- A **hydrocele** is an accumulation of serous fluid in the tunica vaginalis or in a persistent part of the processus vaginalis in the cord.
- A **hematocele** is an accumulation of blood in the tunica vaginalis and results from rupture of testicular blood vessels after trauma to the spermatic cord or testis. Pain associated with trauma to the testis may be referred over the T11–L1 dermatomes.
- A **spermatocele** is a cyst containing sperm that develops in the epididymis just above the testis.
- A **varicocele** results from dilatations of tributaries of the testicular vein in the pampiniform plexus.
  –A varicocele may be caused by defective valves in the pampiniform plexus or by compression of a testicular vein (more often the left) in the abdomen.
  –Varicosities of the pampiniform plexus are observed when the patient is standing and disappear when the patient is lying down. Palpation of the plexus containing the varicocele feels like a bag of worms.

TESTICULAR NEOPLASM

A **malignant neoplasm of the testis** (most commonly a **seminoma**) metastasizes directly to the **lumbar nodes**, distinguishing it from a malignancy of the **scrotum**, which metastasizes initially to **superficial inguinal nodes**. A malignancy is the most common cause of a painless testicular mass.

CRYPTORCHIDISM

*Cryptorchidism is a failure of 1 or both testes to descend completely into the scrotum. The most common location of a cryptorchid testis is in the inguinal canal. If the cryptorchidism is bilateral, the patient may be **sterile**.*

  D. The inguinal canal in females contains the **round ligament of the uterus** and a remnant of the processus vaginalis.
    1. The **round ligament** is the remnant of the **gubernaculum**, which passes through the inguinal canal during fetal life and ends in the labium majus.
    2. A **persistent processus vaginalis** in females forms the **canal of Nuck**.

METASTASIS AND THE ROUND LIGAMENT

*Carcinoma of the fundus of the uterus may metastasize to **superficial inguinal nodes** along lymphatic vessels that course with the round ligament.*

HERNIAS

Inguinal hernias (more common in males) result from a protrusion of an abdominopelvic structure (commonly part of the small intestine) through the anterior abdominal wall in the inguinal region.
- *Inguinal hernias are either indirect or direct; both types may emerge through the superficial inguinal ring and pass superficial to the inguinal ligament.*
- *An **indirect inguinal hernia** is the most common type. Indirect inguinal hernias protrude through the anterior abdominal wall lateral to the inferior epigastric artery and vein, enter the deep inguinal ring, and appear at the superficial ring after traversing the length of the inguinal canal.*
- *Indirect inguinal hernias may follow a persistent processus vaginalis through the inguinal canal, are covered by peritoneum and the 3 fascial layers of the spermatic cord, and may descend into the scrotum. Female patients may develop an indirect inguinal hernia in a canal of Nuck.*
- ***Direct inguinal hernias** protrude through the posterior wall of the inguinal canal medial to the inferior epigastric artery and vein.*
- *Direct inguinal hernias are more likely to tear through transversalis fascia and cremasteric fascia; therefore, they are usually covered only by external spermatic fascia and lie adjacent to the contents of the spermatic cord and the obliterated processus vaginalis at the superficial inguinal ring.*
- ***Femoral hernias** (more common in females) enter the anterior thigh after passing through the femoral ring deep to the inguinal ligament. Femoral hernias have the highest rate of bowel incarceration of any type of hernia.*
- *The **femoral ring**, the site of the protrusion of a femoral hernia, is medial to the femoral vein and lateral to the lacunar ligament, an extension of the inguinal ligament.*
- *After passing deep to the inguinal ligament, femoral hernias may protrude anteriorly through the saphenous hiatus, a fault in the fascia lata that transmits the saphenous vein.*

  V. The **abdominal cavity** contains gastrointestinal structures that are grouped into 3 continuous regions derived from the embryonic foregut, midgut, and hindgut (Figure 4–4). Below the diaphragm, each region differs in peritoneal relationships, blood supply, and innervation.
    A. Peritoneal Relationships
      1. Peritoneal refers to structures that are suspended by a derivative of a dorsal or ventral mesentery in postnatal life.

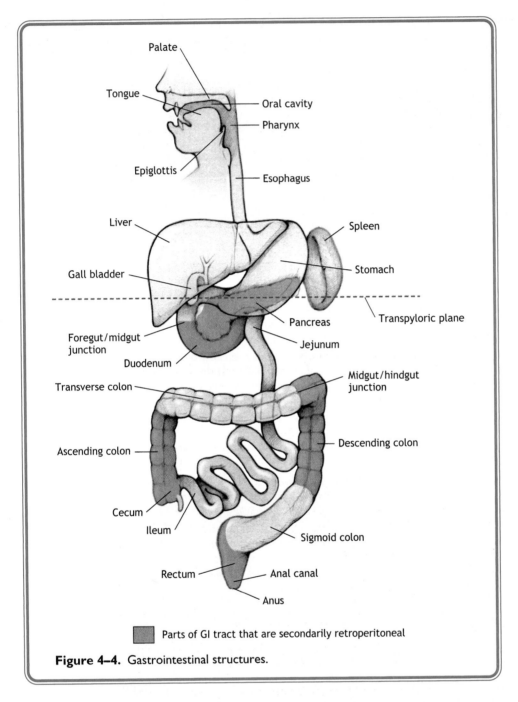

Figure 4–4. Gastrointestinal structures.

2. **Secondarily retroperitoneal** refers to structures that are no longer suspended by a mesentery in postnatal life, are only partially covered by peritoneum, and lie on the posterior wall of the abdominopelvic cavity.

3. **Primarily retroperitoneal** refers to structures that were never suspended by a mesentery, are partially covered by peritoneum, and lie mainly on the posterior wall of the abdominopelvic cavity.

4. A **postnatal derivative** of a mesentery may be called an **omentum** (eg, greater omentum or lesser omentum) or a **ligament** (eg, the gastrosplenic ligament or the hepatoduodenal ligament), or it may be named for the gastrointestinal structure that it suspends (eg, the **transverse mesocolon** [mesentery of the transverse colon] or the **mesoappendix** [mesentery of the appendix]).

B. **Blood Supply**
1. Branches of the **celiac artery** supply gastrointestinal structures in the foregut.
2. Branches of the **superior mesenteric artery** supply gastrointestinal structures in the midgut.
3. Branches of the **inferior mesenteric artery** supply gastrointestinal structures in the hindgut.

C. **Innervation (see Figure 1–5)**
1. Gastrointestinal structures are innervated by parasympathetic, sympathetic, and sensory neurons, which contribute to the enteric nervous system.
2. The **enteric nervous system** is found in the walls of gastrointestinal structures from the esophagus to the anal canal.
3. The enteric nervous system consists of a **myenteric (Auerbach's) plexus** and a **submucosal (Meissner's) plexus**.
   a. Each plexus consists of preganglionic parasympathetic axons, terminal parasympathetic ganglia, postganglionic sympathetic axons, and sensory neurons.
      (1) **Preganglionic parasympathetic axons** to gastrointestinal structures course in the vagus nerves and in the pelvic splanchnic nerves and synapse in the terminal ganglia.
      (2) **Axons of terminal parasympathetic ganglia** in Auerbach's plexus innervate smooth muscle.
      (3) **Axons of terminal ganglia** in Meissner's plexus innervate gastrointestinal glands.
      (4) **Postganglionic sympathetic axons** in these plexuses arise from neuronal cell bodies in the celiac, superior mesenteric, and inferior mesenteric prevertebral ganglia.
   b. The parasympathetic elements of the enteric nervous system inhibit contraction of sphincters, stimulate glandular secretions, and stimulate contraction of gastrointestinal smooth muscle promoting peristalsis.
   c. The sympathetic elements of the enteric nervous system facilitate contraction of smooth muscle sphincters and inhibit peristalsis and glandular secretions.
   d. **Sensory neurons** in enteric plexuses respond to the presence and contents of material in the lumen.
   e. Sensory neurons that respond to pain and innervate gastrointestinal structures course back to the CNS with sympathetic nerves and enter the spinal cord from T5 through L2.

REFERRED PAIN

- *Stimulation of visceral pain fibers* that innervate a gastrointestinal structure results in a dull, aching, poorly localized pain that is referred over the T5 through L1 dermatomes.
- The *sites of referred pain* generally correspond to the spinal cord segments that provide the sympathetic innervation to the affected gastrointestinal structure.

COLICKY PAIN

- A *colicky type of pain* is a rhythmic, recurring pain symptomatic of *ileus*, an obstruction of a gastrointestinal structure.
- *Colicky pain* results from recurrent smooth muscle contractions against the obstruction.
- *Colic* is a severe form of colicky pain.
- *Biliary or renal colic* results from recurring smooth muscle contractions against a gallstone lodged in the biliary system or a calculus lodged in the ureter.

> **4. Parietal peritoneum** is sensitive to pain and is innervated by lower intercostal nerves (T7 through T11), the subcostal nerve (T12), the iliohypogastric and ilioinguinal nerves (L1), and branches of the phrenic nerve (C3–5).

INFLAMMATION OF THE PARIETAL PERITONEUM

Irritation or *inflammation of the parietal peritoneum* by an enlarged gastrointestinal structure or by the escape of fluid from a gastrointestinal structure results in sharp, localized pain over the affected area.
- Patients with *inflamed parietal peritoneum* may exhibit *rebound tenderness* and *guarding* over the site of the inflammation.
- *Rebound tenderness* is pain that is elicited after the pressure of palpation over the affected area is removed.
- *Guarding* refers to reflex spasms of abdominal muscles in response to palpation, which may be evident over the region of the inflamed peritoneum.

> **VI.** The **foregut structures** include the pharynx and the esophagus above the diaphragm, the distal part of the esophagus below the diaphragm, the stomach and the duodenum up to the entrance of the bile duct, and the liver and pancreas, which are outgrowths of the gastrointestinal tract. The spleen is not a foregut derivative but uses the blood supply of the foregut (see Figure 4–5).
>> **A.** Foregut structures are **peritoneal** in postnatal life except for the head, neck, and body of the pancreas.
>>> **1.** The stomach, foregut duodenum, liver, biliary apparatus, and spleen are peritoneal and are suspended by a derivative of the dorsal or ventral mesentery or by a derivative of both mesenteries.
>>> **2.** The **celiac artery**, which supplies the foregut, divides within 1 cm of its origin into the **splenic artery**, the **common hepatic artery**, and the **left gastric artery** (Figure 4–5a and b and Figure 4–6a).
>>>> a. The **splenic artery** supplies
>>>>> *(1)* The spleen
>>>>> *(2)* The neck, body, and tail of the pancreas
>>>>> *(3)* The left side of the greater curvature of the stomach (left gastroepiploic artery)
>>>>> *(4)* The fundus of the stomach (short gastric arteries)

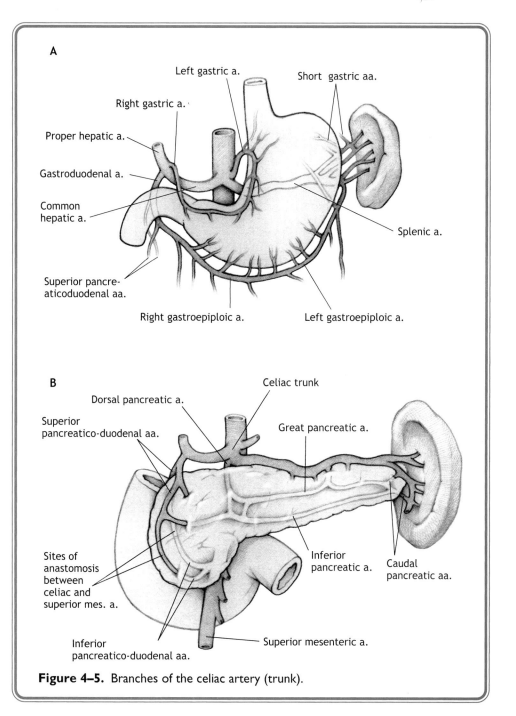

A

Left gastric a.

Short gastric aa.

Right gastric a.

Proper hepatic a.

Gastroduodenal a.

Common
hepatic a.

Splenic a.

Superior pancre-
aticoduodenal aa.

Right gastroepiploic a.

Left gastroepiploic a.

B

Celiac trunk

Dorsal pancreatic a.

Superior
pancreatico-duodenal aa.

Great pancreatic a.

Sites of
anastomosis
between
celiac and
superior mes. a.

Inferior
pancreatic a.

Caudal
pancreatic aa.

Inferior
pancreatico-duodenal aa.

Superior mesenteric a.

Figure 4–5. Branches of the celiac artery (trunk).

b. The **left gastric artery** supplies
       (1) Most of the lesser curvature of the stomach
       (2) The abdominal part of the esophagus
    c. The **common hepatic artery** divides into the proper hepatic artery and
       the gastroduodenal artery.
       (1) The **proper hepatic artery** supplies part of the lesser curvature (right
           gastric artery) and branches into right and left hepatic arteries, which
           supply the liver, and a cystic artery, which supplies the gallbladder.
           The cystic artery is a branch of the right hepatic artery.
       (2) The **gastroduodenal artery** supplies the foregut part of the duode-
           num, the superior part of the head of the pancreas (anterior and poste-
           rior superior pancreaticoduodenal arteries), and the right side of the
           greater curvature of the stomach (right gastroepiploic artery).
 B. The foregut is innervated by parasympathetic and sympathetic axons and vis-
    ceral sensory fibers.
    1. The vagus nerves and neurons in terminal ganglia provide the parasympa-
       thetic innervation.
    2. Preganglionic sympathetic axons in the **lower thoracic splanchnic nerves**
       (greater splanchnic nerves) from the T5–9 spinal cord segments and neurons
       in the celiac ganglia provide the sympathetic innervation.
    3. **Visceral pain fibers** course back to the spinal cord in the greater splanchnic
       nerves. Foregut pain is referred over the T5–9 dermatomes in the epigastric
       and hypochondriac regions.
    4. Sensations other than pain are carried back to the central nervous system
       (CNS) in the vagus nerves.
 C. The **esophagus** enters the abdomen by passing through the right crus of the di-
    aphragm at the level of the T10 vertebra. It consists of smooth muscle, which
    forms a physiologic sphincter near its point of entry into the stomach.

DISORDERS OF THE ESOPHAGUS

Gastroesophageal reflux, or heartburn, *may result from an incompetent lower esophageal sphincter.*
Patients with gastroesophageal reflux complain of substernal burning that is worse when lying down.
In **achalasia** *of the esophagus, the smooth muscle sphincter of the esophagus fails to relax.*
• *Patients have difficulty swallowing liquids and solids. They also have a dilated esophagus and experi-*
  ence abnormal contractions of the smooth muscle of the esophagus proximal to the affected segment.
• *Achalasia is similar to* **Hirschsprung's disease** *in that both may be caused by an absence of terminal*
  parasympathetic ganglia.
An **epiphrenic diverticulum** *may develop just superior to the lower esophageal sphincter.*
• *These diverticula are false or pulsion diverticula that do not consist of all of the layers of the esophagus.*
• **False diverticula** *are most common in the sigmoid colon.*

    D. The **stomach** is found in the epigastric region and consists of the cardia, fundus,
       body, and the pyloric part.
       1. The stomach receives the esophagus at the cardia.
       2. It is continuous with the duodenum at the pyloric sphincter.
       3. It has a concave **lesser curvature** and a convex **greater curvature**.
       4. The stomach is **peritoneal** and is suspended by the lesser omentum, a deriva-
           tive of ventral mesentery, which extends from the lesser curvature to the liver
           (Figure 4–6a, b, c; Figure 4–7a and b).

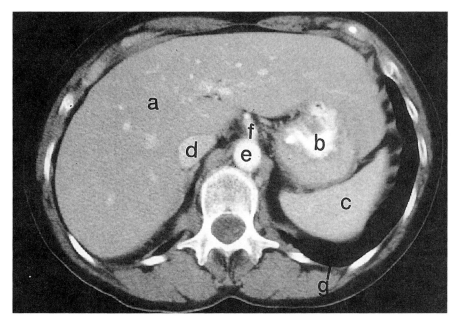

Figure 4–6a. CT images of foregut structures. a. Liver, b. Stomach, c. Spleen, d. Inferior vena cava, e. Abdominal aorta, f. Celiac trunk, g. Costo diaphragmatic recess.

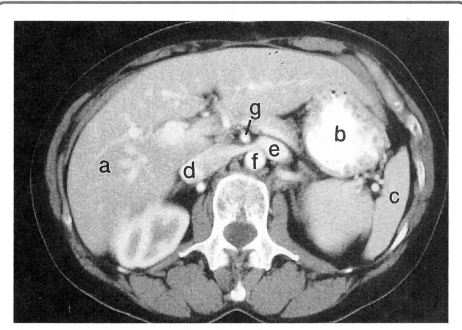

Figure 4–6b. a. Liver, b. Stomach, c. Spleen, d. Inferior vena cava, e. Left renal vein, f. Abdominal aorta, g. Superior mesenteric artery.

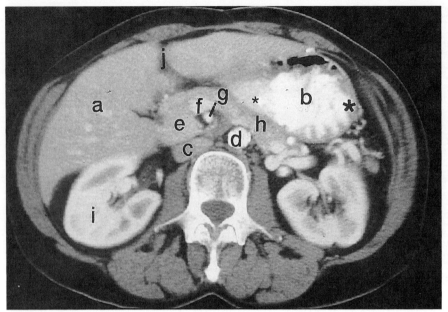

Figure 4–6c. a. Right lobe of the liver, b. Stomach, c. Inferior vena cava, d. Abdominal aorta, e. Head of pancreas, f. Hepatic portal vein, g. Superior mesenteric artery, h. Body of spleen. Large asterisk: Greater curvature of stomach. Small asterisk: Lesser curvature of stomach, i. Kidney, j. Lig. Venosum.

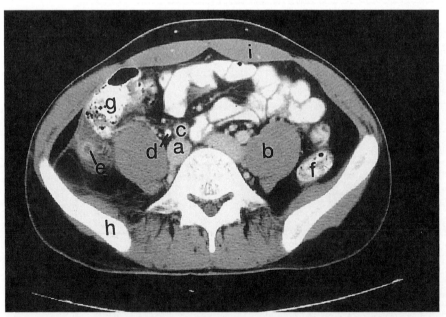

Figure 4–6d. a. Common iliac vein, b. Psoas muscle, c. Common iliac artery, d. Ureter, e. Inflamed appendix, f. Descending colon, g. Ascending colon, h. Ilium, i. Rectus abdominis muscle.

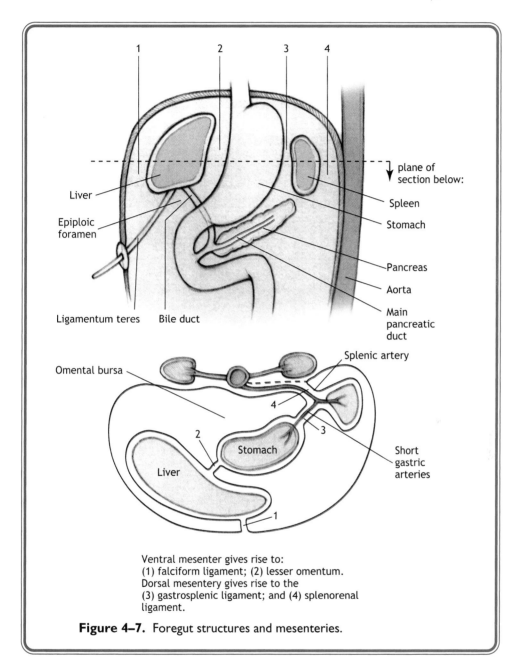

1
2
3
4

plane of
section below:

Liver

Epiploic
foramen

Spleen

Stomach

Pancreas

Aorta

Main
pancreatic
duct

Ligamentum teres   Bile duct

Splenic artery

Omental bursa

4

2

3

Stomach

Short
gastric
arteries

Liver

1

Ventral mesenter gives rise to:
(1) falciform ligament; (2) lesser omentum.
Dorsal mesentery gives rise to the
(3) gastrosplenic ligament; and (4) splenorenal
ligament.

Figure 4–7. Foregut structures and mesenteries.

a. The **hepatoduodenal** and **hepatogastric ligaments** are the parts of the lesser omentum.

b. The **omental bursa,** or lesser peritoneal cavity, lies posterior to the lesser omentum and the stomach and is formed as a result of the rotation of foregut structures during fetal life.

c. The **omental bursa** communicates with the greater peritoneal cavity through the **epiploic foramen** (Figure 4–7a).

d. The **epiploic foramen** is bounded anteriorly by the hepatoduodenal ligament, which contains the hepatic portal vein, the proper hepatic artery, and the bile duct.

e. The **epiploic foramen** is bounded posteriorly by the inferior vena cava, superiorly by the caudate lobe of the liver, and inferiorly by the duodenum.

5. The stomach is also suspended by the **greater omentum**, a derivative of dorsal mesentery that extends from the greater curvature to encircle the spleen before reflecting back onto the dorsal body wall.

a. The **gastrocolic ligament** is part of the greater omentum, which extends from the greater curvature of the stomach to the transverse colon.

b. The **gastrosplenic ligament** extends from the greater curvature to the spleen.

c. The splenorenal ligament extends from the spleen to the dorsal body wall.

SURGICAL ACCESS TO THE OMENTAL BURSA

Surgical access to the omental bursa may be obtained by incising the *lesser omentum*, the *gastrocolic ligament*, or the *gastrosplenic ligament*.

- The *right and left gastric arteries*, which course in the lesser omentum, would have to be avoided in a surgical approach.
- The *middle colic artery*, which courses in the gastrocolic ligament, would have to be avoided in a surgical approach.
- *Short gastric arteries* and the left gastroepiploic artery, which course in the gastrosplenic ligament, would have to be avoided in a surgical approach.

CARCINOMAS OF THE STOMACH

Carcinomas of the stomach commonly develop in the *pyloric part* and metastasize to the *cisterna chyli* and through the *thoracic duct* to the left brachiocephalic vein.

- An enlarged *left supraclavicular node of Virchow* may act as a *sentinel node* for *gastric carcinoma*.
- Carcinomas of the stomach that metastasize to the ovaries are known as *Krukenberg tumors*.

E. The **foregut part of the duodenum** begins at the level of the L1 vertebra distal to the pyloric sphincter and includes a superior part and a descending part up to the site of entrance of the bile duct.

1. The **foregut part of the duodenum** is **peritoneal** and is suspended by the **hepatoduodenal ligament** (part of the lesser omentum). The hepatoduodenal ligament contains the proper hepatic artery, the bile duct, and the portal vein.

2. The foregut part of the duodenum is crossed posteriorly by the gastroduodenal artery and the bile duct.

3. It is known from contrast imaging procedures as the **duodenal ampulla or the duodenal cap.**

CELIAC ARTERY, OCCLUSION AND COLLATERAL CIRCULATION, AND EFFECTS OF ULCERS

*In an **occlusion of the celiac artery** at is origin from the abdominal aorta, collateral circulation may develop in the **head of the pancreas** by way of anastomoses between the **pancreaticoduodenal branches of both the superior mesenteric** and the gastroduodenal arteries.*

*Three branches of the **celiac circulation** may be subject to **erosion** if an ulcer penetrates the posterior wall of the stomach or the posterior wall of the duodenum.*

- *The **splenic artery** may be subject to erosion by the contents of a penetrating ulcer of the **posterior wall** of the stomach.*
- *The **left gastric artery** may be subject to erosion by the contents of a penetrating ulcer of the **lesser curvature** of the stomach.*
- *The **gastroduodenal artery** may be subject to erosion by the contents of a penetrating ulcer of the **posterior wall** of the first part of the duodenum.*
- *Patients with a penetrating ulcer may have **pain referred to the shoulder**, which occurs when air escapes through the ulcer and stimulates the peritoneum covering the inferior aspect of the diaphragm.*

*The contents of a **penetrating ulcer** of the posterior wall of the stomach or the duodenum may enter the **omental bursa**. The fluid contents from an ulcer may pass through the **epiploic foramen** into the **subhepatic recess (Morison's pouch)**, the part of the greater peritoneal cavity situated between the posterior aspect of the liver and the right kidney.*

   F. The **spleen** (Figures 4–4 through 4–8) is situated in the left hypochondriac region.
     1. It functions as a lymphatic organ and a filter for red blood cells.
     2. The spleen is peritoneal and is suspended by the gastrosplenic and splenorenal ligaments.
     3. It is protected posterolaterally by ribs 9 through 12.
     4. The **splenic artery**, which reaches the spleen by coursing posterior to the stomach and superior to the pancreas, supplies the spleen.
     5. The spleen is drained by the splenic vein, which courses posterior to the body and tail of the pancreas.

FRACTURED RIBS AND THE SPLEEN

*A **fractured 9th, 10th, or 11th rib on the left** may **lacerate the spleen**. The spleen bleeds profusely when lacerated and is usually removed.*

   G. The **liver** is situated mainly in the right hypochondriac region and extends into the epigastric region (Figure 4–6a, b, c and Figure 4–9).
     1. The liver functions as an exocrine gland, producing and secreting bile; bile is released into the biliary duct system.
     2. It also functions as an endocrine gland, producing plasma proteins and releasing glucose and lipoproteins into the hepatic sinusoids.
     3. The liver processes nutrients, regulates blood glucose levels, detoxifies drugs and toxins, and stores triglycerides and vitamin A.
     4. It is **peritoneal** and is suspended by derivatives of ventral mesentery, the falciform ligament, the coronary ligaments, and the lesser omentum.
       a. The **lesser omentum** extends from the lesser curvature of the stomach to the liver.
       b. The **falciform ligament** extends from the liver to the ventral body wall, ends inferiorly at the umbilicus, and contains the round ligament of the liver.

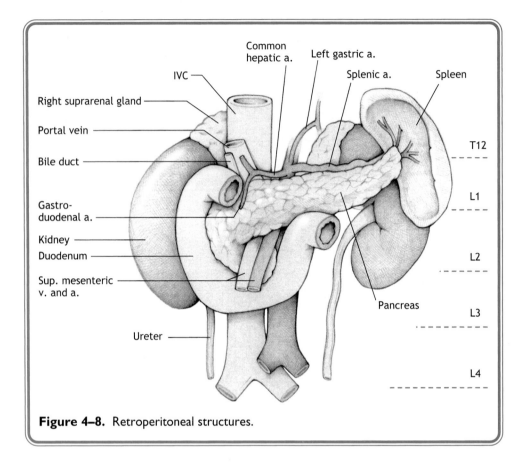

Figure 4–8. Retroperitoneal structures.

5. The **liver** consists of **4 anatomic lobes**: the **left lobe, right lobe, quadrate lobe,** and **caudate lobe** (see Figure 4–9). The fissure containing the round ligament of the liver, a remnant of the **fetal umbilical vein**, and the fissure containing the **ligamentum venosum**, a remnant of the fetal ductus venosus, separate the anatomic right and left lobes.

6. The livers consists of **2 functional lobes** of approximately equal size: the **left lobe** and **the right lobe** (see Figure 4–9).
   a. A line interconnecting the gallbladder and the inferior vena cava generally separates the functional left lobe from the functional right lobe.
   b. The **functional left lobe** includes the left lobe, the quadrate lobe, and part of the caudate lobe; the **functional right lobe** includes the right lobe and part of the caudate lobe.
   c. Each functional lobe has a separate arterial blood supply, venous drainage, and biliary drainage.

7. The liver receives 70% of its blood (poorly oxygenated) from the **portal vein** and 30% (well oxygenated) from the **proper hepatic artery**.
   a. The proper hepatic artery divides into the right hepatic artery and the left hepatic artery, which supply the right and left functional lobes of the liver, respectively.

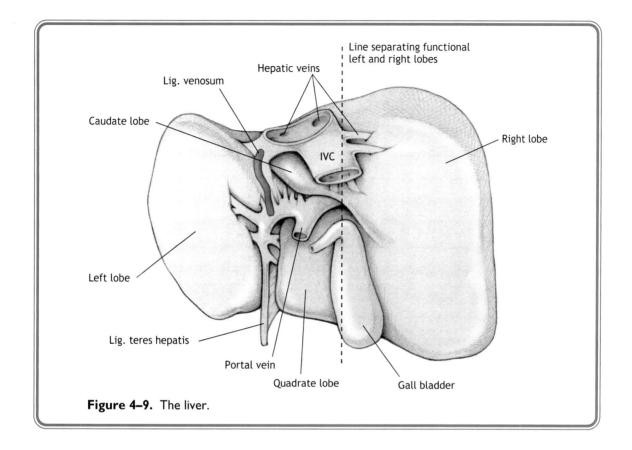

Figure 4–9. The liver.

    b. The hepatic ducts, portal vein, and proper hepatic artery enter or exit the liver at the **porta hepatis** and form the **portal triad** in the **hepatoduodenal ligament**.

   8. **Hepatic veins** drain the liver and empty into the **inferior vena cava** (see Figure 4–9).

CARCINOMAS AND THE LIVER

Primary carcinomas originating in abdominopelvic structures below the diaphragm commonly metastasize to the liver.

   H. The **biliary ducts** collect bile produced by the liver and deliver the bile to the **gallbladder** for storage and to the duodenum for digestion of fats.

    1. The **right and left common hepatic ducts** drain bile from the right and left functional lobes of the liver, respectively, and unite to form the common hepatic duct.

    2. The **common hepatic duct** joins the cystic duct to form the bile duct.

    3. The **bile duct** passes posterior to the duodenum and through the head of the pancreas, joins with the main pancreatic duct to form the **hepatopancreatic ampulla**, and enters the second part of the duodenum at the **major duodenal papilla** (Figure 4–7a).

I. The **gallbladder** stores and concentrates bile produced and secreted by the liver.
1. The gallbladder receives bile from the **cystic duct** and releases bile into the **cystic duct**.
   a. Bile passes through the **cystic duct** and the **bile duct**.
   b. Bile is released into the duodenum in response to the presence of cholecystokinin, which is secreted in response to the presence of fats.
2. The gallbladder is supplied by the **cystic artery**, which is usually a branch of the right hepatic artery.

ANATOMIC VARIATIONS IN BILIARY STRUCTURES

*Anatomic variations in the cystic artery, the cystic duct, and the common hepatic duct, which form the boundaries of the **cystohepatic triangle of Calot**, must be taken into account in surgical procedures involving the gallbladder.*

GALLSTONES

Gallstones may become lodged in the biliary ducts or in the gallbladder.
- *The **hepatopancreatic ampulla**, a narrow point in the biliary duct system, is a common site of an **impacted gallstone**. Patients exhibit referred pain in the epigastric region.*
- *A stone blocking the cystic duct may cause enlargement of the gallbladder. Patients may exhibit **biliary colic**, which is severe, colicky pain that begins in the epigastric region but moves to a point where the ninth costal cartilage intersects the lateral border of the rectus sheath.*
- *An **inflamed gallbladder** may adhere to the duodenum and develop a fistula, permitting a gallstone to pass into the duodenum. The gallstone may become lodged at the ileocecal junction, forming a **gallstone ileus**.*

J. The pancreas is an **exocrine gland** that functions to produce bicarbonates and digestive enzymes that are released by way of a main or an accessory pancreatic duct into the lumen of the duodenum (Figures 4–6c, 4–7a, 4–8).
1. The pancreas is an **endocrine gland** that produces glucagon, insulin, somatostatin, and pancreatic polypeptide.
2. It consists of a head, a neck, a body, and a tail.
   a. The **head, neck, and body** are secondarily **retroperitoneal;** the **tail is peritoneal** and is located in the **splenorenal ligament.**
   b. The head of the pancreas includes the **uncinate process,** is enclosed by the duodenum, and is supplied by pancreaticoduodenal branches of the gastroduodenal (celiac circulation) and superior mesenteric arteries.
   c. The **splenic vein** and the **superior mesenteric vein** join to form the **portal vein** posterior to the neck of the pancreas.
   d. The superior mesenteric vein and artery pass posterior to the neck of the pancreas and anterior to the uncinate process.

PANCREATIC ADENOCARCINOMAS

*Adenocarcinomas of the pancreas commonly develop in the **head of the pancreas** and may result in compression of the bile duct and the main pancreatic duct.*
- *Patients with adenocarcinoma of the pancreas have **epigastric pain** that frequently radiates to the back.*
- *Patients with an **obstructed bile duct** experience obstructive jaundice, which results from a backup of bile pigments and a yellow staining of peripheral tissues.*

- *If the main pancreatic duct is obstructed, the pancreas may become inflamed; patients with acute pan-creatitis may experience a **localized ileus** in the duodenum adjacent to the area of inflammation.*

VII. The **midgut structures** include the distal three-fourths of the duodenum inferior to the entrance of the bile duct; the jejunum; the ileum; the cecum, appendix, and the ascending colon; and the proximal two-thirds of the transverse colon (see Figure 4–4).

A. Midgut structures are **secondarily retroperitoneal or peritoneal** in postnatal life; peritoneal parts of the midgut are suspended by derivatives of a dorsal mesentery.

 1. The midgut part of the duodenum and the ascending colon are secondarily retroperitoneal.

 2. The jejunum, ileum, and proximal two-thirds of the transverse colon are peritoneal.

B. Branches of the **superior mesenteric artery** supply the midgut (Figures 4–6b and c, 4–10, and 4–11).

 1. **Anterior and posterior inferior pancreaticoduodenal** branches supply the midgut part of the duodenum and the inferior part of the head and uncinate process of the pancreas.

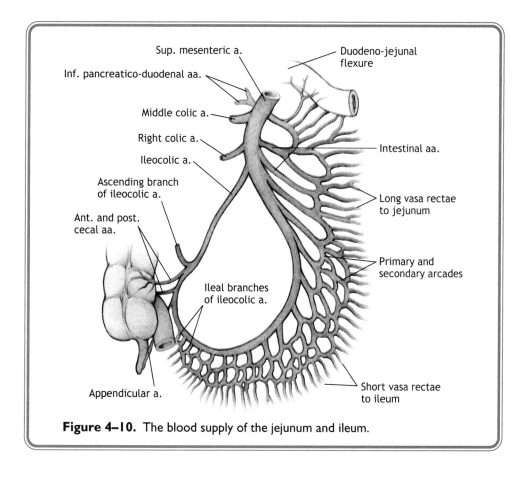

Figure 4–10. The blood supply of the jejunum and ileum.

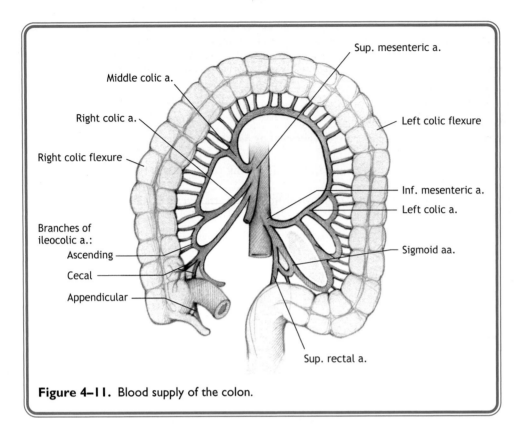

Figure 4–11. Blood supply of the colon.

> 2. **Intestinal branches** that supply the jejunum form simple arcades and long vasa recta.
> 3. **Intestinal branches** that supply the ileum form complex arcades and short vasa recta (Figure 4–11).
> 4. The **ileocolic, right colic, and middle colic arteries** also supply the midgut parts of the colon.
>> a. The **ileocolic artery** supplies the distal part of the ileum, the appendix, the cecum, and the proximal part of the ascending colon.
>> b. The **right colic artery** supplies the ascending colon.
>> c. The **middle colic artery** supplies the proximal two-thirds of the transverse colon (Figure 4–12).
> C. The midgut is innervated by parasympathetic and sympathetic axons and visceral sensory fibers.
>> 1. Preganglionic parasympathetic axons in the **vagus nerves** and neurons in terminal ganglia provide the parasympathetic innervation.
>> 2. Preganglionic sympathetic axons in the **lower thoracic splanchnic nerves** (greater, lesser, and least splanchnic nerves) from the T5 through the T12 spinal cord segments and neurons in the celiac and the superior mesenteric ganglia provide the sympathetic innervation.

3. **Visceral pain fibers** course back to the spinal cord in lower thoracic splanchnic nerves.
      a. **Pain from the small intestine** part of the midgut tends to be referred over the T5 through 10 dermatomes in the epigastric region and umbilical region.
      b. **Pain from the cecum, appendix, ascending colon, and proximal two-thirds of the transverse colon** tends to be referred over the T10 through T12 dermatomes in the umbilical, hypogastric, and lumbar regions.
   4. Visceral sensations other than pain from the midgut are carried back to the CNS in the vagus nerves.
D. The **midgut part of the duodenum** begins in the second part of the duodenum distal to the entrance of the bile duct (see Figures 4–4 and 4–7a).
   1. The midgut part of the duodenum is **secondarily retroperitoneal**.
   2. It includes the descending (or second) part of the duodenum distal to the entrance of the bile duct, the horizontal (third) part, and the ascending (fourth) part.
      a. The **descending part** extends inferiorly to the level to the L3 vertebra.
      b. The **horizontal part** passes from right to left anterior to the body of the L3 vertebra and is crossed anteriorly by the superior mesenteric artery and vein.
      c. The **ascending part** extends superiorly from the left of the L3 vertebra to the left of the L2 vertebra at the duodenojejunal junction.
      d. The descending, horizontal, and ascending parts of the duodenum enclose the head of the pancreas.

DUODENAL COMPRESSION

*The **superior mesenteric vessels** may compress the horizontal part of the duodenum. Patients experience epigastric or umbilical pain, nausea after a meal, and bilious vomiting.*

E. The **jejunum** begins to the left of the L2 vertebra at the duodenojejunal junction and is generally situated in the upper left quadrant (see Figure 4–4).
   1. The jejunum is **peritoneal** throughout its length and is suspended by the mesentery proper.
      a. The **ligament of Treitz**, or suspensory ligament of the duodenojejunal junction, is a fibromuscular band that supports the duodenojejunal flexure.
      b. The ligament of Treitz contains muscle from the right crus of diaphragm.

GASTROINTESTINAL BLEEDING AND THE LIGAMENT OF TREITZ

***Hematemesis**, a vomiting of blood, commonly results from bleeding into the lumen of the esophagus, stomach, or duodenum proximal to the ligament of Treitz. Hematemesis is commonly caused by a duodenal ulcer, a gastric ulcer, or esophageal varices.*
***Hematochezia,** blood in the stool, usually results from bleeding into the lumen of the jejunum, ileum, colon, or rectum distal to the ligament of Treitz.*
***Melena** refers to black, tarry stools that contain blood altered by gastric secretions. In **melanemesis** there is a "coffee-ground" appearance of the vomitus.*

   2. The jejunum is continuous with the ileum; 40% of the peritoneal part of the small intestine is jejunum and 60% is ileum.
   3. The jejunum has thicker walls, a larger luminal diameter, more prominent plica circulares, and greater vascularity than the ileum.

F. The **ileum** is continuous with the jejunum and ends at the ileocecal junction with the ascending colon (see Figure 4–4).

    1. The ileum has more fat in its mesentery than the jejunum.

    2. It contains **Peyer's patches** in its wall.

MECKEL'S DIVERTICULUM

*An **ileal or Meckel's diverticulum** is a remnant of the **fetal vitelline duct**, which persists in postnatal life as an outpocketing of the ileum.*

- *A **diverticulum** is an outpocketing of a tubular or saccular organ such as the gastrointestinal tract or the bladder. **True diverticula** are protrusions that include of all of the layers of the affected structure; **false or pulsion diverticula** are protrusions that do not contain all tissue layers.*

- *A **Meckel's diverticulum** is a true diverticulum that is located approximately 2 ft from the ileocecal junction, occurs in about 2% of the population, is about 2 in long, and may contain ectopic gastric or pancreatic cells.*

- *Patients may have bleeding associated with ulceration of the ectopic cells, obstruction, and pain that is referred over the area of the umbilical region in the T10 dermatome.*

- *The pain associated with an acute Meckel's diverticulum may mimic referred pain associated with an inflamed vermiform appendix.*

G. The **midgut part of the colon** (see Figure 4–11) is continuous with the ileum at the ileocecal junction.

    1. It includes the **cecum and ascending colon**, which are secondarily retroperitoneal.

    2. The midgut part of the colon includes the proximal two-thirds of the **transverse colon**, which is peritoneal and suspended by the transverse mesocolon.

    3. It is characterized by

        a. A larger luminal diameter than the small intestine.

        b. **Teniae coli**, 3 independent longitudinal bands of smooth muscle, which begin at the base of the appendix and end at the junction between the sigmoid colon and the rectum.

        c. **Haustrations**, or sacculations formed because the teniae coli are shorter than the total length of the colon.

        d. **Omental appendices**, which are globules of fat covered by peritoneum suspended from the teniae coli.

    4. It contains the **vermiform appendix** (see Figures 4–4 and 4–6db).

        a. The appendix is peritoneal and suspended by the mesoappendix and contains lymphatic tissue.

        b. The base of the vermiform appendix arises from the cecum inferior to the ileocecal opening.

        c. The appendix is supplied by the **appendicular artery**, a branch of the ileocolic artery.

        d. The distal part of the appendix is commonly situated in a retrocecal location. Less frequently, it extends inferiorly and medially across the pelvic brim.

APPENDICITIS

*In **appendicitis**, the vermiform appendix may become inflamed as a result of either an obstruction by a stool, which forms a **fecalith** (common in adults), or **hyperplasia** of its lymphatic tissue (common in children).*

- An inflamed appendix (see Figure 4–6d) may stimulate visceral pain fibers, which course back in the lower splanchnic nerves and result in colicky pain referred over the umbilical region.
- In these patients, irritation of the parietal peritoneum may result in pain localized over the base of the appendix (**McBurney's point**). McBurney's point is one-third of the distance along a line between the anterior superior iliac spine and the umbilicus in the lower right quadrant.
- The **iliohypogastric nerve** may be lesioned in an appendectomy procedure; a weakening of the anterior abdominal wall and a direct inguinal hernia may result.

INTESTINAL INTUSSUSCEPTION

In an **intussusception**, part of the small intestine invaginates or telescopes into an adjacent distal segment, the **intussuscipiens**.
- An intussusception may be jejunoileal, ileoileal, or, most commonly, **ileocecal**, where the distal part of the ileum telescopes into the ascending colon.
- An intussusception is more common in children than in adults and may be caused by hyperplasia of lymphatic tissue in Peyer's patches in the wall of the ileum.
- Patients may have an obstructed bowel, right-sided colicky pain, abdominal distension, and hematochezia because the blood supply to the intussuscepted ileum may be compromised.
- The obstruction and vascular compromise associated with an intussusception are similar to those in patients with **sigmoid volvulus**.

VIII. The **hindgut structures** include the distal one-third of the transverse colon, descending colon, sigmoid colon, rectum, and anal canal to the pectinate line (see Figure 4–4).

   A. Hindgut structures are secondarily retroperitoneal or peritoneal in postnatal life; peritoneal parts of the hindgut are suspended by derivatives of a dorsal mesentery.

   1. The descending colon, the rectum, and the anal canal are secondarily retroperitoneal.

   2. The distal one-third of the transverse colon and the sigmoid colon are peritoneal.

   B. **Branches of the inferior mesenteric artery supply the hindgut** (see Figure 4–11).

   1. The **left colic artery** supplies the descending colon and the distal third of the transverse colon.

   2. **Sigmoid arteries** supply the sigmoid colon.

   3. The **superior rectal artery** supplies the rectum and anal canal to the pectinate line.

   4. The **middle rectal artery**, a branch of the internal iliac artery, contributes to the blood supply of the rectum and the anal canal above the pectinate line.

   C. The hindgut is innervated by parasympathetic and sympathetic axons and visceral sensory fibers.

   1. Preganglionic parasympathetic axons in **pelvic splanchnic nerves** that arise from sacral spinal cord segments S2 through S4 and terminal ganglia provide the parasympathetic innervation.

   2. Preganglionic sympathetic axons in the **lower thoracic and lumbar splanchnic nerves** from the T11 to L2 spinal cord segments and neurons in the inferior mesenteric ganglia provide the sympathetic innervation.

3. **Visceral pain fibers** course back to the spinal cord mainly in lower thoracic and lumbar splanchnic nerves.
   a. Pain from the distal third of the transverse colon, descending colon, sigmoid colon, rectum, and anal canal is referred over the T11 through L2 dermatomes in the hypogastric and iliac regions.
   b. Sensations other than pain from the hindgut are also carried back to the CNS in the pelvic splanchnic nerves.

D. The **distal third of the transverse colon** is peritoneal and suspended by the transverse mesocolon.
   1. It is continuous with the descending colon at the splenic or left colic flexure.
   2. The distal third of the transverse colon contains teniae coli, haustrations, and omental appendages.

E. The **descending colon** is secondarily retroperitoneal and descends inferiorly from the **splenic flexure** on the left to the iliac fossa. It contains teniae coli, haustrations, and omental appendages.

F. The **sigmoid colon** is peritoneal and is suspended by the **sigmoid mesocolon.**
   1. It begins in the iliac fossa and is continuous with the rectum at the level of the S3 vertebra.
   2. The sigmoid colon contains teniae coli, haustrations, and omental appendages. The sigmoid-rectal junction marks the end of the teniae coli, the haustrations, and the omental appendices.

SIGMOID VOLVULUS, DIVERTICULOSIS, AND DIVERTICULITIS OF THE SIGMOID COLON

CLINICAL CORRELATION

In **sigmoid volvulus,** the sigmoid colon twists around the sigmoid mesocolon and may become obstructed.
• Patients may experience left-sided colicky pain, abdominal distension, and hematochezia as a result of compromise of the sigmoid arteries.
• Volvulus less commonly involves the cecum.

The sigmoid colon is a common site of **multiple pulsion diverticula.**
• Sigmoid diverticula are false diverticula that form when the mucosa and submucosa herniate through the smooth muscle of the sigmoid colon.
• **Diverticulosis** refers to diverticula that are not inflamed.
• **Diverticulitis** refers to diverticula that are inflamed. If a diverticulum ruptures, the ruptured contents may irritate the parietal peritoneum, resulting in pain that is localized to the left lower quadrant. The signs and symptoms of acute sigmoid diverticulitis are similar to those of acute "left-sided" appendicitis. Patients may have hematochezia.

G. **The rectum** is secondarily retroperitoneal and begins at the level of the S3 vertebra.
   1. The rectum becomes continuous with the anal canal anterior to the tip of the coccyx at the pelvic diaphragm.
   2. The inferior aspect of the rectum is dilated and forms the ampulla, a storage site for feces before defecation.

ISCHEMIC BOWEL INFARCTION

CLINICAL CORRELATION

Common sites of **ischemic bowel infarction** are in the transverse colon near the splenic flexure and in the rectum.

- **Infarction of the transverse colon** occurs between the distal parts of the vascular beds of the middle colic branches of the superior mesenteric and left colic branches of the inferior mesenteric arteries.
- **Infarction of the rectum** occurs between the distal parts of the vascular beds of the superior rectal branches of the inferior mesenteric artery and the middle rectal branches of the internal iliac artery.

HIRSCHSPRUNG'S DISEASE

Hirschsprung's disease is caused by a failure of **neural crest cells** either to migrate into the hindgut or to differentiate into **terminal parasympathetic ganglia** in the walls of the hindgut.
- Patients experience constriction in the affected segment (most often in the rectum), an absence of peristalsis, and a dilated large bowel proximal to the affected segment.
- Hirschsprung's disease is common in patients with **Down syndrome**.

IX. **Venous drainage** of the gastrointestinal structures below the diaphragm is provided by the hepatic portal system (Figures 4–6c and 4–12).
   A. **The Hepatic Portal Vein**
   1. The **hepatic portal vein** receives poorly oxygenated, nutrient-rich blood from venous capillary beds in the walls of gastrointestinal structures.
   2. It delivers that blood to the second capillary bed formed by the **hepatic sinusoids** in the liver.
   3. The **hepatic portal vein** is formed posterior to the neck of the pancreas when the **superior mesenteric vein**, which drains the midgut, unites with the **splenic vein**, which drains part of the foregut.
      a. The **inferior mesenteric vein**, which drains the hindgut, usually empties into the splenic vein.
      b. **Left and right gastric veins**, which drain the stomach, including the lesser curvature and the abdominal part of the esophagus, empty directly into the portal vein.
   B. **Hepatic Veins**
   1. Hepatic veins drain blood from the **hepatic sinusoids**.
   2. They empty into the **inferior vena cava** just below the diaphragm.

CIRRHOSIS AND PORTACAUAL ANASTOMOSES

In **cirrhosis** of the liver, obstruction of the portal vein and **hypertension** in the portal system are caused by destruction of hepatocytes and replacement of hepatocytes by fibrous tissue.
- Patients with cirrhosis of the liver may develop **portal hypertension**, in which venous blood from gastrointestinal structures, which normally enters the liver by way of the portal vein, is forced to flow in the opposite or retrograde direction in tributaries of the portal vein.
- Retrograde flow forces portal venous blood into capillary beds, which also drain into tributaries of the superior or inferior vena cava; **portacaval anastomoses** are established at these sites, permitting portal venous blood to bypass the liver (see Figure 4–12).
- One site of a **portacaval anastomosis** is in the wall of the esophagus at the junction of capillary beds draining into the left gastric (or "coronary") vein (portal system) and the esophageal tributaries of the azygos vein (caval system).
   –**Esophageal varices** are dilated and tortuous veins that develop in the submucosal venous plexus in the walls of the esophagus.
   –Esophageal varices may burst and result in hematemesis.

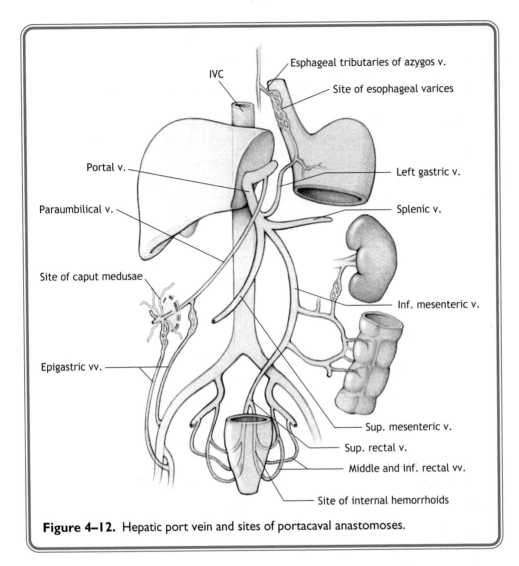

Figure 4–12. Hepatic port vein and sites of portacaval anastomoses.

- *A second site of a portacaval anastomosis is in the wall of the rectum at the junction of the internal rectal plexus, which drains into the superior rectal vein (portal system), and the external rectal plexus, which drains into the middle or inferior rectal veins (caval system).*
  –Internal hemorrhoids are painless protrusions of the anal canal covered by mucosa.
  –Internal hemorrhoids contain dilated veins of the internal rectal venous plexus.
- *A third common site of a portacaval anastomosis is in the anterior abdominal wall at the junction of capillary beds draining into the paraumbilical veins, which course in the falciform ligament (portal system), and the tributaries of the superficial epigastric veins, which drain the anterior abdominal wall (caval system). A caput medusa is a pattern of varicose superficial epigastric veins that radiate away from the umbilicus. Splenomegaly is also a common sign associated with patients with portal hypertension.*

X. The **posterior abdominal wall** contains the kidneys, the ureters, and the adrenal glands.

  A. The **kidneys** are primarily retroperitoneal and are embedded in perirenal and pararenal fat (see Figures 4–6c and 4–8).

    1. The kidneys remove metabolic wastes from the blood and help maintain the blood's ionic balance.

    2. The **left kidney** is superior to the **right kidney** and is situated anterior to the 11th and 12th ribs; the right kidney is situated anterior to the 12th rib.

    3. The kidneys are in contact with the diaphragm and the transversus abdominis, quadratus lumborum, and the psoas major muscles.

    4. The anteromedial border is located at the renal hilus of the kidney and contains the renal artery, renal vein, and renal pelvis, the expanded superior part of the ureter.

      a. At the **hilus**, the renal vein is anterior to the renal artery.

      b. The renal artery is anterior to the renal pelvis.

    5. The kidneys are organized into **cortical and medullary regions**, which contain parts of the uriniferous tubules (the nephrons and collecting ducts).

      a. The **cortex** contains the proximal parts of the uriniferous tubules, the renal corpuscles, and the proximal and distal convoluted portions of the duct system.

      b. The **medulla** consists mainly of renal pyramids, which contain the distal parts of the collecting ducts; at the apex of each pyramid, the collecting ducts open into a minor calyx.

      c. The straight parts of the uriniferous tubules (loops of Henle) extend into **renal columns** situated between the **renal pyramids**.

      d. Two to 4 minor calyces unite to form a major calyx; 3 to 4 major calyces unite to form the renal pelvis, the expanded upper part of the ureter.

    6. The **renal arteries**, which supply the kidneys, arise from the abdominal aorta at the level of the upper part of the L2 vertebra, and renal veins, which drain the kidneys, empty into the inferior vena cava at the same vertebral level.

      a. The renal arteries give rise to **interlobar** arteries that branch into arcuate arteries at the juncture between the medulla and the cortex; arcuate arteries give rise to **interlobular** arteries.

      b. The interlobular arteries give rise to the afferent arterioles of the glomerulus; blood leaves the glomerulus via the efferent arterioles.

INTERLOBULAR ARTERY OCCLUSION

*The interlobular arteries are end arteries. **Occlusion of an interlobular** artery may result in avascular necrosis, leaving a shallow scar in the renal cortex.*

      c. The right renal artery passes posterior to the inferior vena cava, and the left renal vein passes anterior to the abdominal aorta just inferior to the origin of the superior mesenteric artery (Figure 4–10a).

COMPRESSION OF THE URETER

*The proximal part of the ureter may be compressed by an **aberrant renal artery**.*
* *An aberrant renal artery commonly arises inferior to the renal artery and passes anterior to the origin of the ureter. It may cause **hydronephrosis** of the renal pelvis.*

• *A male patient may have a **varicocele** (more common on the left) resulting from compression of the left renal vein by an aneurysm of the superior mesenteric artery near the origin of the artery from the abdominal aorta.*

    7. The kidneys are innervated by preganglionic sympathetic axons in the **lower thoracic or lumbar splanchnic nerves** from the T11 to L2 spinal cord segments, which synapse mainly in **aorticorenal ganglia**.
        a. Postganglionic sympathetic axons from the aorticorenal ganglia are distributed mainly to the renal arteries in the kidney.
        b. Postganglionic sympathetic axons control arterial blood flow to the glomeruli.
  B. The **ureters** are tubes containing smooth muscle that arise from the renal pelvis of each kidney at about the level of the L2 vertebra (see Figures 4–6d and 4–8).
    1. The ureters descend inferiorly on the anterior surface of the psoas major muscle, cross the pelvic brim, and enter the posterolateral aspect of the bladder. At the pelvic brim, the ureter lies between the external and the internal iliac arteries.
    2. **Several arteries supply the ureters**.
        a. The upper third is supplied by the **renal artery**.
        b. The middle third is supplied by the **common iliac artery**.
        c. The distal third is supplied by the **superior vesical artery**.

KIDNEY TRANSPLANTATION

In **kidney transplants**, only the upper part of the ureter, which is supplied by the renal artery, is transplanted with renal vessels and the kidney. In most transplants, the kidneys are placed in the pelvis, where the upper part of the ureter is attached to the bladder, and the renal artery is joined to the external iliac artery.

CALCULI OF THE URETER

A **calculus** may become lodged at 1 of 3 narrow points in the ureter and result in **hydronephrosis** proximal to the site of the blockage. The **narrow points** in the ureter are at the origin of the ureter from the renal pelvis, where the ureter crosses the pelvic brim, and the point at which the ureter enters the bladder.

    3. The ureters are innervated by preganglionic sympathetic axons in the lower thoracic or lumbar splanchnic nerves from the T11 to L2 spinal cord segments, which synapse mainly in aorticorenal ganglia.
        a. Axons from the aorticorenal ganglia innervate smooth muscle in the ureters.
        b. They stimulate contraction of the smooth muscle of the ureters and movement of urine from the renal pelvis to the bladder.

RENAL COLIC

Renal colic is a severe type of **colicky pain** that results from distention of a ureter by a calculus and is referred over the T11 through L2 dermatomes. The pain may radiate from the back above the iliac crest, through the inguinal region, and into the scrotum or labium majus.

  C. The **adrenal glands** are endocrine organs that are primarily retroperitoneal and situated near the superior pole of each kidney.
    1. Each adrenal gland consists of a cortex that contains 3 zones.
        a. The zona glomerulosa cells produce and secrete aldosterone.
        b. The zona fasciculata produces and secretes glucocorticoids.
        c. The zona reticularis produces and secretes reproductive steroids.
    2. The **medulla** of the adrenal gland contains chromaffin cells, which are derived from **neural crest cells** that produce and secrete mostly epinephrine and some

norepinephrine. Chromaffin cells are innervated by preganglionic sympathetic axons, which reach the adrenal medulla in lower thoracic splanchnic nerves.

3. The adrenal glands are supplied by adrenal arteries, which branch from the renal arteries, the aorta, and the inferior phrenic arteries.

4. The adrenal glands are drained by a single vein.

    a. The right adrenal vein drains directly into the inferior vena cava.

    b. The left adrenal vein drains into the left renal vein.

D. The **abdominal aorta** (Figure 4–13) begins at the aortic hiatus at the level of the T12 vertebra and descends inferiorly to the left of the midline. It ends by bifurcating into the common iliac arteries at the level of the L4 vertebra (see Figure 4–6d).

AORTIC ANEURYSM

*A common site for an **aneurysm of the** aorta is just proximal to the **bifurcation of the aorta** at the level of the L4 vertebra. Patients have a pulsating painless mass in the midline.*

1. The abdominal aorta gives rise to 3 unpaired arteries that supply derivatives of the foregut, midgut, and the hindgut.

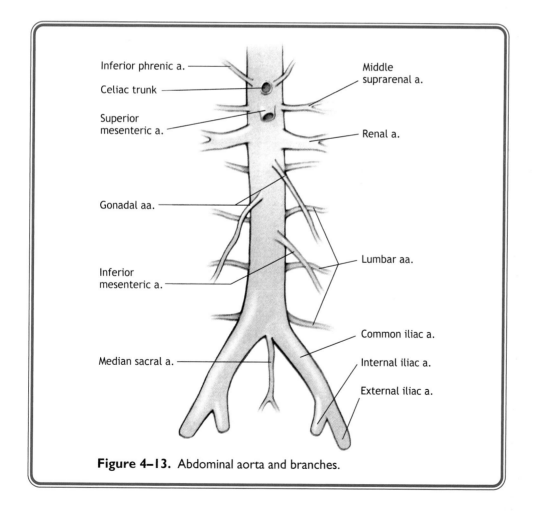

Figure 4–13. Abdominal aorta and branches.

a. The **celiac artery** arises from the aorta at about the level of the **T12 vertebra.**

b. The **superior mesenteric artery** arises from the aorta at about the level of the **L1 vertebra.**

c. The **inferior mesenteric artery** arises from the aorta at about the level of the **L3 vertebra.**

2. It gives rise to several pairs of arteries.

a. The **renal arteries** arise from the aorta at about the level of the **L2 vertebra.**

b. The **gonadal arteries** arise from the aorta between the origins of the renal arteries and the inferior mesenteric artery between the **L2 and the L3 vertebrae.**

c. A pair of inferior phrenic arteries supplies the diaphragm, and several pairs of lumbar arteries supply the body wall.

3. The abdominal aorta gives rise to multiple arteries that supply the adrenal glands.

E. The **inferior vena cava** (Figure 4–14) is formed at about the level of the L5 vertebra by the union of the common iliac veins. It ascends just to the right of the midline.

1. It receives a pair of veins from the kidneys.

a. On the right, the renal, adrenal, and gonadal veins drain directly into the inferior vena cava.

b. On the left, only the **left renal vein** drains directly into the inferior vena cava; the left gonadal and the left adrenal veins drain into the left renal vein. The left renal vein crosses the anterior aspect of the aorta just inferior to the origin of the superior mesenteric artery.

COMPRESSION OF THE LEFT RENAL VEIN AND A VARICOCELE

CLINICAL CORRELATION

*The **left renal vein may be compressed** by an **aneurysm of the superior mesenteric artery** as the vein crosses anterior to the aorta. Patients with compression of the left renal vein may have renal and adrenal hypertension on the left, and, in males, a varicocele on the left.*

2. The inferior vena cava receives 2 or 3 hepatic veins from the liver that empty into the inferior vena cava just inferior to the caval hiatus of the diaphragm.

XI. The **thoracoabdominal diaphragm** consists of skeletal muscle that separates the thoracic cavity from the abdominopelvic cavity (Figure 4–15).

A. The diaphragm attaches to the xiphoid process, the lower 6 ribs, and the bodies of the first 2 or 3 lumbar vertebrae. The right and left crura of the diaphragm attach to the bodies of the upper 3 lumbar vertebrae.

B. The diaphragm is innervated by motor fibers in the **phrenic nerves**, which contain ventral rami from the C3, C4, and C5 spinal cord segments.

1. **Sensory fibers in the phrenic nerves** innervate most of the pleura on the superior aspect of the diaphragm, the mediastinal pleura, and most of the peritoneum on the inferior aspect of the diaphragm.

2. **Sensory fibers in intercostal nerves** innervate the pleura and peritoneum in the peripheral parts of the diaphragm.

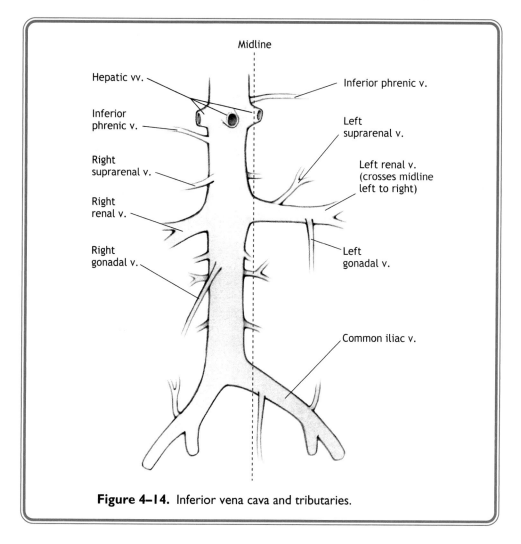

Figure 4–14. Inferior vena cava and tributaries.

Labels in figure:

Midline

Hepatic vv.

Inferior phrenic v.

Inferior phrenic v.

Left suprarenal v.

Right suprarenal v.

Left renal v. (crosses midline left to right)

Right renal v.

Right gonadal v.

Left gonadal v.

Common iliac v.

PERITONEAL IRRITATION

Irritation of the peritoneum *covering the diaphragm may result in pain being referred to the **C3, C4, or C5 dermatomes in the region of the neck** and shoulder.*

    C. The diaphragm contains **3 hiatuses** that transmit neural, vascular, or visceral structures (see Figure 4–13).

        1. The **caval hiatus** is at the level of the body of the T8 vertebra and transmits the inferior vena cava and the right phrenic nerve.

        2. The **esophageal hiatus** is in the right crus of the diaphragm, is at the level of the body of the T10 vertebra, and transmits the esophagus and the anterior and posterior vagal trunks.

        3. The **aortic hiatus** is located between the right and left crura at the level of the T12 vertebra and transmits the aorta, the azygos vein, and the thoracic duct.

CLINICAL CORRELATION

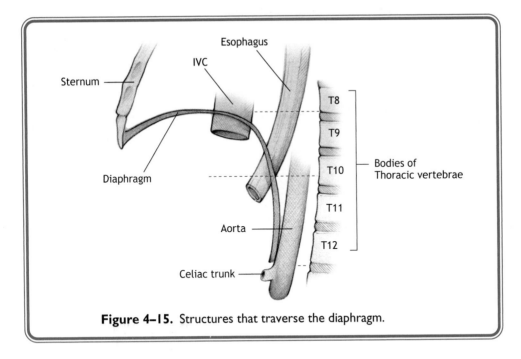

Figure 4–15. Structures that traverse the diaphragm.

    a. The thoracic duct originates from the cisterna chyli, which is situated just inferior and to the right of the aortic hiatus.

    b. Virtually all lymph from structures in the perineum, pelvis, and lower limbs drain into lumbar nodes and then by right and left lumbar trunks into the cisterna chyli.

    c. An intestinal lymphatic trunk that drains gastrointestinal structures below the diaphragm joins with the lumbar trunks at the cisterna chyli.

4. The **sternocostal triangles** situated on either side of the xiphisternal joint transmit the superior epigastric artery and vein.

5. The **right and left crura** of the diaphragm transmit the greater, lesser, and lower splanchnic nerves that arise from thoracic spinal cord segments from T5–12 and synapse in the prevertebral (celiac, superior mesenteric, or aorti-corenal) ganglia below the diaphragm.

DIAPHRAGMATIC HERNIA

Diaphragmatic hernias occur at 1 of 4 locations.

• A **congenital diaphragmatic hernia** may occur in the left posterolateral part of the diaphragm just anterior to the quadratus lumborum.

• A **hiatal hernia** may occur at the esophageal hiatus. In a **sliding hiatal hernia**, the most common type of hiatal hernia, the abdominal part of the esophagus and part of the stomach herniate through the esophageal hiatus into the mediastinum. **Gastroesophageal reflux** or **heartburn** may result from a sliding hiatal hernia.

• A **paraesophageal hiatal hernia** may also occur at the esophageal hiatus where the fundus or body of the stomach herniates into the mediastinum adjacent to the esophagus.

• A **retrosternal diaphragmatic hernia** may occur in the anterior part of the diaphragm adjacent to the xiphisternal joint.

CLINICAL PROBLEMS

In a male patient, an aneurysm of the superior mesenteric artery has compressed structures that cross the midline inferior to the origin of the artery from the abdominal aorta.

1. What sign/symptom might the patient have?

    A. A spermatocele in the left scrotum

    B. A hydrocele in the left scrotum

    C. Internal hemorrhoids

    D. Pain and vomiting after a meal

    E. Hydronephrosis on the right

Your patient experiences portal hypertension resulting from cirrhosis of the liver.

2. Between which of the following pairs of veins might collateral circulation develop to enable blood to bypass the liver?

    A. Left gastric vein/azygos vein

    B. Hepatic vein/inferior vena cava

    C. Inferior epigastric vein/superficial epigastric vein

    D. Middle rectal vein/inferior rectal vein

    E. Splenic vein/superior mesenteric vein

Your patient has carcinoma of the pancreas, which compresses structures coursing through it.

3. You might expect the patient to have all of the following except:

    A. Splenomegaly

    B. Internal hemorrhoids

    C. Pancreatitis

    D. Caput medusa

    E. Reduced arterial pressure in the right colic artery

A 56-year-old man experiences persistent gastroesophageal reflux. Diagnostic imaging reveals that the patient has a sliding hiatal hernia.

4. What other structure might be compressed by the hernia that courses through the same opening in the diaphragm?

    A. Phrenic nerve

    B. Greater splanchnic nerve

    C. Thoracic duct

    D. Azygos vein

    E. Vagal trunks

Your patient suffers from a fractured rib on the posterolateral aspect of the thorax on the left. The spleen is ruptured and has to be removed surgically, along with a significant part of the splenic artery.

5. Removal of the distal part of the splenic artery might disrupt the blood supply to what structure?

   A. Left adrenal gland

   B. Greater curvature of the stomach

   C. Head of the pancreas

   D. Third part of the duodenum

   E. Left ureter

A 38-year-old banker with a history of heartburn suddenly experiences excruciating pain in the epigastric region of the abdomen. Surgery is performed immediately on admission to the emergency room. There is evidence of an ulcer, which has ruptured through the posterior wall of the duodenum.

6. What blood vessel might be subject to erosion?

   A. Common hepatic

   B. Left gastric

   C. Splenic

   D. Superior mesenteric

   E. Gastroduodenal

A surgeon incises the lesser omentum of a patient to gain access to the omental bursa to remove the ulcerated material.

7. What blood vessel has to be avoided in this approach?

   A. Short gastric arteries

   B. Left gastroepiploic artery

   C. Gastroduodenal artery

   D. Proper hepatic artery

   E. Celiac artery

A 23-year-old secretary in good health suddenly doubles over in pain emanating from the area of the umbilicus. The secretary feels warm and uneasy and has no appetite. That night the pain becomes sharper and more localized and seems to have moved to the lower right abdominal region. The secretary calls her family physician, who arranges for an ambulance to bring the patient to the hospital.

8. Which nerves most likely carried the painful sensations into the central nervous system that resulted in pain referred to the area of the umbilicus?

   A. Vagus nerves

   B. Lower thoracic splanchnic nerves

   C. Pelvic splanchnic nerves

   D. Iliohypogastric nerves

   E. Genitofemoral nerves

An infant is born with defects in the migration and differentiation of neural crest cells.

9. Which of the following gastrointestinal conditions might be evident?
   A. Sigmoid volvulus
   B. Intussusception
   C. Achalasia of the esophagus
   D. Meckel's diverticulum
   E. Diverticulitis

10. In the patient in Question 9, what ganglia failed to develop?
    A. Paravertebral ganglia
    B. Celiac ganglia
    C. Terminal ganglia
    D. Prevertebral ganglia
    E. Dorsal root ganglia

A male construction worker comes to the outpatient clinic after work complaining of groin pain and a bulge that appears when he lifts heavy objects. When the patient lies down, the bulge disappears. A diagnosis of an indirect hernia is made.

11. Which of the following is characteristic of this type of hernia?
    A. The hernia has protruded through the anterior abdominal wall medial to the inferior epigastric blood vessels.
    B. The hernia may tear through the internal spermatic and cremasteric fasciae that cover the spermatic cord.
    C. The hernia may pass posterior to the inguinal ligament into the anterior thigh.
    D. The hernia may pass through the superficial inguinal ring but not through the deep inguinal ring.
    E. The hernia is most likely covered by all of the same layers that cover the spermatic cord.

A surgeon is performing an exploratory inspection of a patient's abdomen.

12. What anatomic feature permits the surgeon to distinguish the ileum from other parts of the small intestine?
    A. It is the only retroperitoneal part of the small intestine.
    B. It has more fat in its mesentery than the jejunum.
    C. It has thicker walls than the jejunum.
    D. It has wider teniae coli than the jejunum.
    E. It is supplied by intestinal branches of the left colic artery.

A 34-year-old man is diagnosed with a tumor of the adrenal medulla, on the left, and surgery is indicated to remove the entire gland. During surgery, the vasculature to the left adrenal (suprarenal) gland is ligated.

13. Which of the following must be ligated?

   A. Two left adrenal veins that drain into the left gonadal vein and the IVC, respectively.

   B. Three adrenal veins that drain into the left renal, the IVC, and the splenic vein, respectively.

   C. Three adrenal veins that drain into the left renal, the inferior mesenteric, and the lumbar veins, respectively.

   D. A single left adrenal vein that drains directly into the inferior vena cava (IVC).

   E. A single left adrenal vein that drains into the left renal vein.

A woman develops a blockage of the inferior mesenteric artery at its origin from the aorta that results in decreased blood supply to the transverse colon.

14. Which of the following vessels would most likely provide collateral circulation to the affected part of the transverse colon?

   A. Ileocolic artery

   B. Intestinal artery

   C. Left colic artery

   D. Middle colic artery

   E. Right colic artery

The celiac artery has become occluded at the point where it arises from the abdominal aorta.

15. Which of the following is a potential site of an anastomosis between branches of the celiac circulation and the superior mesenteric circulation?

   A. Right gastric artery with the left gastric artery

   B. Superior pancreaticoduodenal arteries with inferior pancreaticoduodenal arteries

   C. Right gastroepiploic artery with the left gastroepiploic artery

   D. Short gastric arteries with the splenic artery

   E. Right gastric artery with the gastroduodenal artery

A 56-year-old male is undergoing a transplant of the right kidney, right renal vein, and right renal artery.

16. Which of the following is correct regarding this procedure?

   A. The ureter from the renal pelvis to the point where it crosses the pelvic brim will be transplanted since the right renal artery supplies this segment.

   B. The proximal part of the right renal vein must not be removed, because it typically drains the right suprarenal gland.

   C. The right renal artery will be ligated at the point where it arises from the aorta between the testicular arteries and the inferior mesenteric artery.

   D. The IVC must be avoided because the right renal artery passes posterior to this vessel in its course from the aorta to the right kidney.

ANSWERS

1. The answer is D. The aneurysm has compressed the third part of the duodenum and possibly the left renal vein, both of which cross the aorta inferior to the origin of the superior mesenteric artery. The patient may exhibit pain and vomiting after a meal and a varicocele in the left spermatic cord.

2. The answer is A. Esophageal varices develop as a result of the portacaval anastomosis between the esophageal branches of the left gastric vein and tributaries of the azygos vein.

3. The answer is D. Caput medusa may develop in patients with liver cirrhosis but not in those with pancreatic carcinoma.

4. The answer is E. The vagal trunks traverse the esophageal hiatus, the site of a hiatal hernia.

5. The answer is B. The splenic artery supplies the spleen, greater curvature and fundus of the stomach, and the body and tail of the pancreas.

6. The answer is E. The gastroduodenal artery courses immediately posterior to the posterior part of the duodenum.

7. The answer is D. The proper hepatic artery courses in the hepatoduodenal ligament, which is part of the lesser omentum.

8. The answer is B. The lower thoracic splanchnic nerves carry visceral pain from an inflamed appendix back into the CNS. Only sympathetic nerves carry visceral pain.

9. The answer is C. Achalasia results from an absence of ganglion cells in the lower part of the esophagus.

10. The answer is C. Terminal ganglia are parasympathetic ganglia that develop in the walls of gastrointestinal structures.

11. The answer is E. The hernia is most likely covered by all of the same layers that cover the spermatic cord. Choices A, B, and D are characteristic of a direct inguinal hernia; choice C is characteristic of a femoral hernia.

12. The answer is B. The ileum has more fat in its mesentery than the jejunum but has thinner walls and is less vascular.

13. The answer is E. A single left adrenal vein drains into the left renal vein.

14. The answer is D. Branches of the middle colic artery (superior mesenteric) anastomose with branches of the left colic artery (inferior mesenteric).

15. The correct answer is B. The superior pancreaticoduodenal arteries (gastroduodenal) anastomose with inferior pancreaticoduodenal arteries (SMA).

16. The correct answer is D. The IVC must be avoided because the right renal artery passes posterior to this vessel in its course from the aorta to the right kidney.

I. The **bony pelvis** is formed by 2 hip, or innominate bones which articulate anteriorly at the **pubic symphysis** and posteriorly with the sacrum at the **sacroiliac joints**.

A. Hip Bones
1. Each hip bone consists of 3 bones: the **ilium, ischium,** and **pubis.**
2. These bones are fused together at the **acetabulum.**

B. Greater Pelvis
1. The **greater pelvis** is bounded largely by the iliac fossae, the sacrum above the promontory, and the anterior abdominal wall.
2. In females, the greater pelvis is shallower than in males.
3. The greater pelvis contains the iliacus muscle, which largely fills the iliac fossa and forms the posterolateral muscular wall of the greater pelvis.

C. Lesser Pelvis
1. The **lesser pelvis** is formed by the hip bones, the sacrum below the promontory, and the coccyx.
2. In females, the lesser pelvis is shallower and wider than in males.
3. The lesser pelvis contains the obturator internus and the piriformis muscles, which form the lateral walls of the lesser pelvis.

D. Pelvic Inlet (Figure 5–1)
1. The **pelvic inlet** separates the greater pelvis and the lesser pelvis and is formed by the superior aspect of the pubic symphysis, the pelvic brim, and the sacral promontory.
2. The pelvic inlet tends to be oval in females and heart shaped in males.

E. Pelvic Outlet
1. The **pelvic outlet** is a diamond-shaped opening; its borders include the inferior aspect of the pubic symphysis, the ischiopubic rami, the ischial tuberosities, the sacrospinous ligaments, and the tip of the coccyx.
2. The **ischiopubic rami** unite at the pubic symphysis to form the **subpubic angle**. The subpubic angle is approximately 90 degrees in females and approximately 60 degrees in males.

PELVIC DIMENSIONS AND VAGINAL DELIVERY

*The **three smallest dimensions of the pelvis** may be evaluated in anticipation of a pregnancy and a **vaginal delivery**.*

CLINICAL CORRELATION

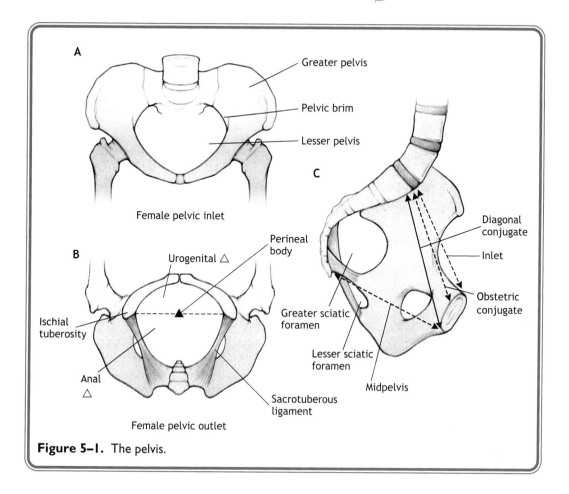

Figure 5–1. The pelvis.

- At the **pelvic inlet,** the smallest anteroposterior dimension is that of the **obstetric conjugate** (Figure 5–1), which extends from the midpoint of the pubic symphysis to the sacral promontory.
- The **midpelvis** lies on a plane extending from the inferior aspect of the pubic symphysis through the ischial spines to the sacrum. The smallest dimension within this plane, and the smallest overall dimension of the lesser pelvis, is the distance between the **ischial spines.**
- The size of the **pelvic outlet** can be estimated by measuring the distance between **ischial tuberosities** or the size of the **subpubic angle.**

    3. The pelvic outlet is subdivided into a **urogenital triangle** and an **anal triangle** by a coronal line extending between the ischial tuberosities.
        a. The **urogenital triangle** is situated directly behind the pubic symphysis; its posterior border corresponds to the line drawn between the ischial tuberosities.
        b. The **anal triangle** is posterior to the urogenital triangle; its lateral borders are the sacrotuberous ligaments, and its anterior border corresponds to the line drawn between the ischial tuberosities.

4. The **perineal body** (central tendon of the perineum) lies in the midline between the ischial tuberosities.

II. The **pelvic diaphragm** is shaped like a funnel, consists of skeletal muscles and fascia, and serves as a boundary between the pelvis and the perineum (Figures 5–2 to 5–4, Table 5–1).

A. The **base of the pelvic diaphragm** attaches to the bony and fascial walls of the lesser pelvis just below the pelvic brim.

B. The **apex of the pelvic diaphragm** projects posteriorly and inferiorly into the anal triangle.

C. The bladder, prostate, and proximal part of the urethra, the uterus and upper part of the vagina, and the rectum are situated in the pelvis above the pelvic diaphragm.

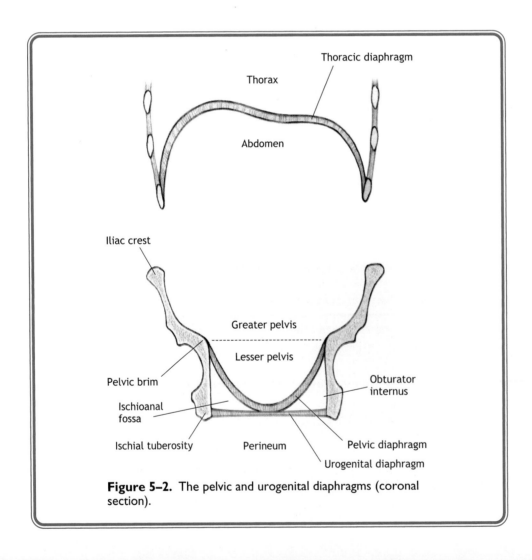

Figure 5–2. The pelvic and urogenital diaphragms (coronal section).

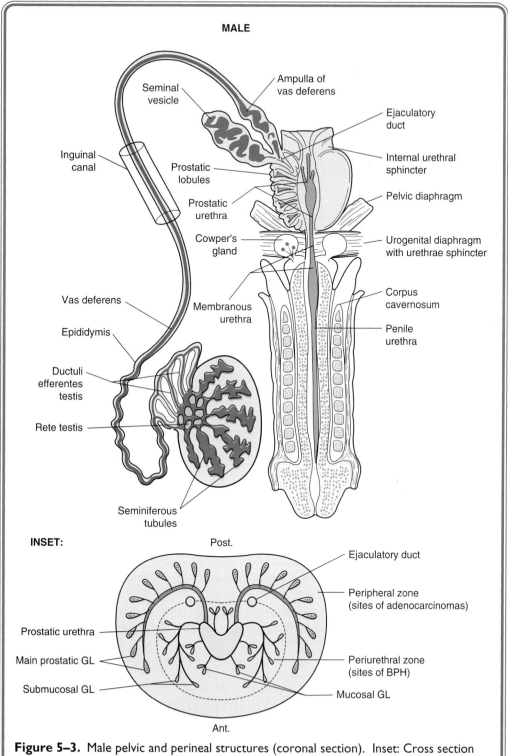

MALE

Ampulla of vas deferens

Seminal vesicle

Ejaculatory duct

Inguinal canal

Internal urethral sphincter

Prostatic lobules

Prostatic urethra

Pelvic diaphragm

Cowper's gland

Urogenital diaphragm with urethrae sphincter

Vas deferens

Corpus cavernosum

Epididymis

Penile urethra

Membranous urethra

Ductuli efferentes testis

Rete testis

Seminiferous tubules

INSET:

Post.

Ejaculatory duct

Peripheral zone (sites of adenocarcinomas)

Prostatic urethra

Periurethral zone (sites of BPH)

Main prostatic GL

Submucosal GL

Mucosal GL

Ant.

Figure 5–3. Male pelvic and perineal structures (coronal section). Inset: Cross section through the prostate. GL, gland; BPH, benign prostatic hyperplasia.

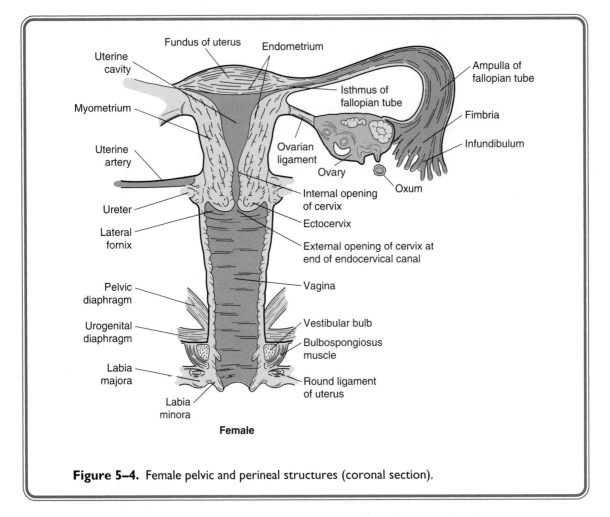

Figure 5–4. Female pelvic and perineal structures (coronal section).

D. The **urogenital hiatus** is a gap in the anterior part of the pelvic diaphragm and permits passage of the urethra in both sexes and the vagina from the lesser pelvis to the perineum.

E. The **pelvic diaphragm** consists of **2 skeletal muscles** that are named from their attachments to the pubis, the ilium, and the ischium.

1. The **levator ani** forms the anterior two-thirds of the pelvic diaphragm. It consists of the **pubococcygeus, puborectalis,** and **iliococcygeus** muscles.

a. The **pubococcygeus** muscle encircles the vagina (females), prostate (males), and anorectal junction in both sexes and attaches to the perineal body.

b. The **puborectalis** bounds the urogenital hiatus and forms part of the external anal sphincter. When the puborectalis contracts, the anorectal junction is pulled anteriorly, contributing to fecal continence.

c. The **iliococcygeus** muscle arises mostly from the tendinous arch, a thickening of the fascia covering the obturator internus muscle, and attaches to the anococcygeal raphe between the anal canal and the coccyx.

Table 5–1. Muscles of Pelvic Diaphragm, Urogenital Diaphragm, and Superfical Perineal Space

Pelvic Diaphragm: Actions	Muscles Involved	Innervation
Support pelvic viscera Contract to increase intra-abdominal pressure	1. Levator ani A. Pubococcygeus Puborectalis	Ventral rami of S3 and S4 and pudendal n.
Relax during micturition, defecation, parturition	B. Iliococcygeus 2. "Ischio" coccygeus	Ventral rami of S3 and S4 Ventral rami of S4 and S5
Compresses anal canal in maintaining continence	External anal sphincter	Inferior rectal branch of pudendal (S2, 3, 4)
Urogenital Diaphragm: Action	**Muscles Involved**	**Innervation**
Compresses urethra	Sphincter urethrae	Perineal branch of pudendal (S2, 3, 4)
Supports perineal body	Deep transverse perineus	Perineal branch of pudendal (S2, 3, 4)
Superficial Perineal Space: Action	**Muscles Involved**	**Innervation**
Compresses penile urethra	Bulbospongiosus	Perineal branch of pudendal (S2, 3, 4)
Maintains erection	Ischiocavernosus	Perineal branch of pudendal (S2, 3, 4)
Supports perineal body	Superficial transverse perineus	Perineal branch of pudendal (S2, 3, 4)

2. The **(ischio) coccygeus muscle** forms the posterior third of the pelvic diaphragm. The coccygeus muscle attaches laterally to the ischial spine and sacrospinous ligament and medially to the coccyx and sacrum.
F. The **pelvic diaphragm** is innervated directly by ventral rami of the S2 and S3 spinal nerves and by branches of the pudendal nerve.
G. Contractions of the **pelvic diaphragm** increase intra-abdominal pressure.
H. It is enclosed by the ischioanal fossa (Figure 5–2).
   1. The **ischioanal fossa** is filled with fat and connective tissue and is situated between the pelvic diaphragm and the obturator internus muscle and fascia.
   2. The **pudendal nerve** and the internal pudendal artery and vein course in the lateral wall of the ischioanal fossa through the pudendal canal. The fascia covering the obturator internus muscle forms the pudendal canal.

PELVIC DIAPHRAGM WEAKNESS

Weakness of the pelvic diaphragm may result in the *prolapse* of part of the uterus into the vagina or herniation of the bladder or rectum into the vagina.

- *In patients with **uterine prolapse,** the cervix, isthmus, and body of the uterus protrude into the superior aspect of the vagina. The patient may experience bleeding and discharge into the vagina.*
- *In patients with a **cystocele,** the bladder herniates into the upper part of the anterior wall of the vagina. The patient may experience urinary problems.*
- *In patients with a **rectocele,** the rectum herniates into the lower part of the posterior wall of the vagina. The patient may have difficulty in defecation. **Kegel exercises** strengthen the pelvic diaphragm, in particular the pubococcygeus muscles, to prevent prolapse or herniation of pelvic viscera.*

   III. Pelvic viscera include the urinary bladder and urethra, anal canal, and male and female reproductive structures.

      A. The Urinary Bladder and Urethra (Figures 5–5 and 5–7)

         1. The **urinary bladder** is primarily retroperitoneal and consists of a bag that stores urine produced in the kidneys.

            a. The superior aspect of the bladder is covered by peritoneum; the peritoneum reflects onto the anterior aspect of the uterus in females, forming a **vesicouterine pouch,** and onto the anterior aspect of the rectum in males, forming a **rectovesical pouch.**

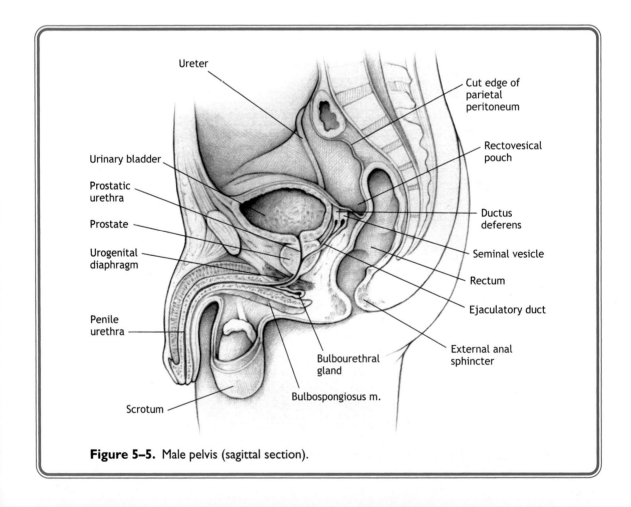

Figure 5–5. Male pelvis (sagittal section).

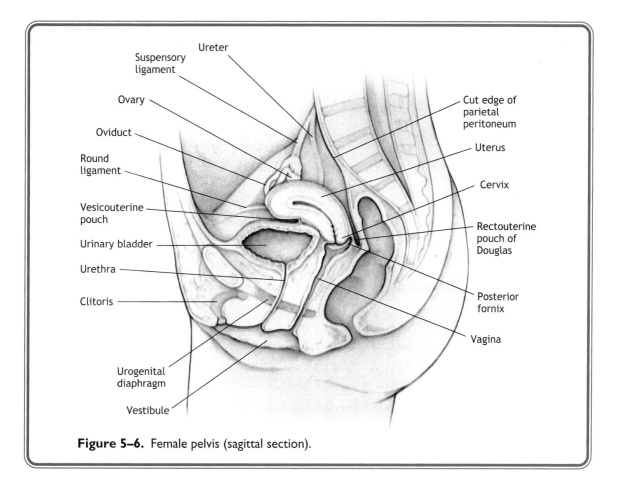

Figure 5–6. Female pelvis (sagittal section).

  b. When empty, the bladder lies below and posterior to the pubic bone. When full, it may expand superiorly above the pubic bone, elevating its covering layer of peritoneum with it.
  c. The **trigone** is on the inner posterior wall of the bladder. The **ureters** open into the bladder at the lateral angles of the trigone. The urethra opens inferiorly at the apex of the trigone.
2. The **bladder** contains the detrusor muscle and has 2 sphincter muscles that contract to maintain urinary continence.
  a. The **detrusor muscle** is a smooth muscle that contracts to aid in the passage of urine from the bladder into the urethra during micturition.
  b. The **sphincter vesicae,** or internal urethral sphincter (Figure 5–3), is a smooth muscle that encircles the urethra at the junction of the bladder and the urethra.
  c. The **sphincter urethrae,** or external urethral sphincter, is a skeletal muscle that encircles the urethra in the urogenital diaphragm.
3. The **male urethra** is approximately 12 cm in length and has 3 parts (Figure 5–3).

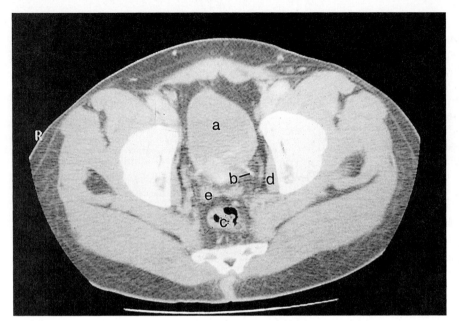

Figure 5–7a. CT scan of male pelvis through the bladder. a. Bladder, b. Ureter, c. Rectum, d. Obturator internus muscle, e. Seminal vesicle.

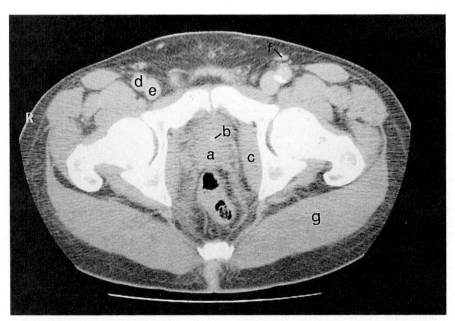

Figure 5–7b. CT scan of male pelvis through the prostate. a. Prostate, b. Prostatic urethra, c. Obturator internus muscle, d. Femoral artery, e. Femoral vein, f. Spermatic cord, g. Gluteal muscles.

a. The **prostatic urethra** arises from the bladder and passes through the prostate and the urogenital hiatus of the pelvic diaphragm to reach the urogenital diaphragm. The sphincter vesicae muscle is situated between the bladder and the prostate.

     b. The **membranous urethra** passes through the sphincter urethrae muscle in the urogenital diaphragm.

     c. The **penile urethra** turns sharply anteriorly inferior to the urogenital diaphragm and traverses the bulb and corpus spongiosum of the penis.

  4. The **female urethra** (Figure 5–6) is approximately 5 cm in length and does not course through the clitoris. The proximal part of the urethra arises from the bladder, passes through the urogenital hiatus and the urogenital diaphragm, and opens into the vestibule just anterior to the opening of the vagina.

B. The Anal Canal (Figures 5–5 and 5–6)

  1. The **anal canal** is in the midline of the anal triangle between the perineal body and coccyx. It is joined to the coccyx by the **anococcygeal ligament**.

  2. It contains **anal columns,** which are ridges of mucosa that contain branches of the superior rectal artery and vein. The anorectal junction is at the superior end of the anal columns.

  3. The **anal canal** is divided into 2 parts by the pectinate line, which is located at the inferior ends of the anal columns.

  4. Above the **pectinate line:**

     a. The anal canal is lined by mucosa.

     b. The arterial blood supply of the anal canal is provided mainly by the superior rectal artery.

     c. The venous drainage is provided by the internal rectal venous plexus, which drains mainly into the superior rectal vein; the superior rectal drains into the inferior mesenteric vein, a tributary of the hepatic portal system.

     d. Lymph from the anal canal drains initially to internal iliac nodes.

     e. Sensory innervation of the mucosa is provided by fibers that course with autonomic nerves; the mucosa is insensitive to pain.

INTERNAL HEMORRHOIDS

Internal hemorrhoids *are painless protrusions of the anal canal covered by mucosa. They contain dilated veins of the internal rectal venous plexus.*

  5. Below the **pectinate line:**

     a. The anal canal is lined by skin.

     b. The arterial blood supply of the anal canal is provided by the inferior rectal artery, which is a branch of the internal pudendal artery.

     c. The venous drainage is provided by the external rectal venous plexus, which drains into the inferior rectal vein; the inferior rectal vein drains into the internal iliac vein by way of the internal pudendal vein, a tributary of the caval system.

     d. Lymph from the anal canal drains initially to superficial inguinal nodes.

     e. Sensory innervation of the skin is provided by the inferior rectal branches of the pudendal nerve; the skin is sensitive to touch, pain, and temperature.

EXTERNAL HEMORRHOIDS

External hemorrhoids are painful enlargements covered by skin that contain dilated veins of the external rectal venous plexus.

    6. The anal canal **has internal and external sphincter muscles,** which contract to maintain fecal continence.
        a. The **internal anal sphincter** consists of smooth muscle derived from the inner circular muscle layer of the rectum.
            (1) The internal anal sphincter is under involuntary control.
            (2) The internal anal sphincter is innervated by sympathetic nerves, which facilitate contraction of the sphincter, and parasympathetic nerves, which facilitate relaxation.
        b. The **external anal sphincter** consists of skeletal muscle derived in part from the puborectalis muscle. The external anal sphincter attaches to the skin and to the perineal body.
            (1) The external anal sphincter is under voluntary control.
            (2) The external anal sphincter is innervated by the inferior rectal branches of the pudendal nerve.
  C. **Male Reproductive Structures**
    1. The **ductus deferens** arises from the **epididymis** adjacent to the testis, traverses the inguinal canal (Figure 5–3), and then descends into the pelvis to reach the posterior aspect of the urinary bladder. The ductus deferens transports sperm, which are produced in the seminiferous tubules of the testis.
    2. The **seminal vesicles** are situated on the posterolateral aspect of the bladder just lateral to the ductus deferens.
        a. The **seminal vesicles** secrete a fructose-rich component of seminal fluid.
        b. Seminal vesicle secretions contribute to sperm metabolism and make up a major component of seminal fluid.
    3. An **ejaculatory duct** is formed when a ductus deferens joins with a duct of the seminal vesicle just above the prostate.
        a. The ejaculatory ducts transport sperm and seminal fluid through the prostate.
        b. The ejaculatory ducts open onto the surface of the seminal colliculus in the posterior wall of the prostatic urethra.
    4. The **prostate** surrounds the prostatic urethra and consists of glandular tissue and smooth muscle.
        a. The **periurethral zone** of the prostate surrounds the prostatic urethra and contains mucosal and submucosal glands; their secretions lubricate the urethra.
        b. The **peripheral zone** of the prostate contains the main prostatic glands.
        c. The **main prostatic glands** secrete a milky component of seminal fluid that contains prostatic acid phosphatase and prostate-specific antigen and contributes to sperm activation and motility..
        d. The **ducts of the prostatic glands** open into the prostatic sinuses on either side of the seminal colliculus in the prostatic urethra.
        e. The **ejaculatory ducts** pass through the posterior part of the prostate between the periurethral zone and the peripheral zone.

BENIGN PROSTATIC HYPERPLASIA

Benign prostatic hyperplasia (BPH) commonly occurs in the **periurethral zone** of the prostate and may result in obstruction of the prostatic urethra.
- *Obstruction of the urethra may impede urinary flow and result in incomplete emptying of the bladder.*
- *Patients with BPH have difficulty initiating urination and have an increased need to urinate.*

ADENOCARCINOMA OF THE PROSTATE

Adenocarcinomas commonly develop in the peripheral part of the prostate in the main prostatic glands (most commonly in the posterior part).
- *Urine flow may be altered and patients may pass blood in the urine.*
- *Prostate adenocarcinomas commonly metastasize to bones of the pelvis, to the bodies of vertebrae, or to ribs.*
- *Patients have elevated blood levels of prostatic acid phosphatase and prostate-specific antigen.*

   D. Female Reproductive Structures
1. The **uterus** is a pear-shaped organ situated between the bladder and the rectum. It contains smooth muscle and consists of a fundus, a body, an isthmus, and a cervix (Figures 5–4, 5–6, and 5–8).
   a. The **fundus** is superior to the openings of the uterine tubes, which are located at the junction of the body and the fundus.
   b. The **body** of the uterus begins below the uterine tubes and includes the isthmus.
   c. The **cervix** (endocervical canal) begins inferior to the isthmus at the internal os and continues to the external os, which opens into the upper end of the vagina.
   d. The **uterus** is typically **anteverted** in 50% of females (folded anteriorly at the junction of the cervix and the vagina) and lies on the superior aspect of the bladder (Figure 5–6).
   e. The uterus is **retroverted** (folded posteriorly) in approximately 25% of females and in a **midposition** in 25% of females.

BICORNUATE UTERUS

A bicornuate uterus results when there is a lack of complete fusion of the paramesonephric (Mullerian) ducts (Figure 5–8b). In a common form of this congenital defect, the uterus has two horns that fuse near the cervix, and open into the vagina.

    f. The uterus facilitates movement of sperm from the cervix to the opening of the uterine tubes, provides a protective and nourishing environment for the developing embryo and fetus, and contracts to push the mature fetus and the placenta into the vagina during labor.
2. The **uterine tubes** consist of an intramural part, an isthmus, an ampulla, and an infundibulum.
   a. The **intramural part** lies inside the uterine wall between the fundus and the body.
   b. The **isthmus** has a thick muscular wall and extends from the uterine wall to the ampulla.
   c. The **ampulla** is dilated and is the longest segment of the uterine tube.

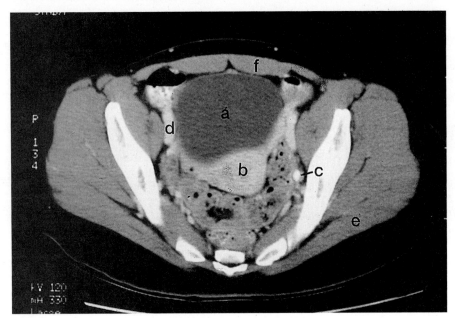

Figure 5–8a. CT scan of cross section through female pelvis. a. Bladder, b. Body of uterus, c. Ureter, d. External iliac, e. Gluteus maximus, f. Rectus abdominis.

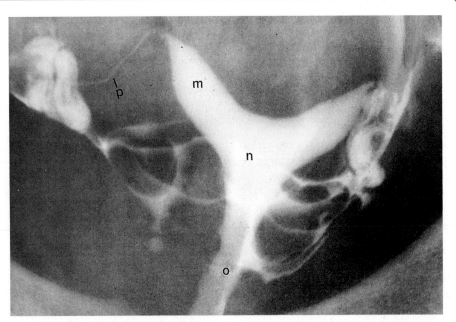

Figure 5–8b. Bicornuate uterus containing contrast media. m. Horn of bicornuate uterus, n. Body of uterus, o. Vagina, p. Uterine tube.

d. The **infundibulum** is the distal segment of the uterine tube. The distal end of the infundibulum terminates in fimbriae and contains the ostium, which opens into the peritoneal cavity.

e. The uterine tubes facilitate transport of sperm from the uterus and transport of an ovum to the uterus.

3. The **ovaries** lie adjacent to the infundibulum of the uterine tube. They contain the developing ovarian follicles.

OVULATION

At **ovulation,** a secondary oocyte is extruded through the wall of the ovary and into the peritoneal cavity (Figure 5–4).

- The ostium of the infundibulum of the uterine tube is in close proximity to the ovary.
- The fimbriae of the infundibulum trap the ovulated ovum and move it to the ostium.

FERTILIZATION

In **fertilization,** a sperm fuses with a secondary oocyte; the most common site of fertilization is in the ampulla of the uterine tube.

ECTOPIC IMPLANTATION

An **ectopic implantation** also most commonly occurs in the **ampulla** of the uterine tube.

- Patients with an ectopic tubal implantation may have amenorrhea, vaginal bleeding, and abdominal pain. The abdominal pain may mimic that of appendicitis.
- Patients with a ruptured tubal implantation have intraperitoneal bleeding, which may accumulate in the rectouterine pouch of Douglas.
- The most common cause of a tubal implantation is **pelvic inflammatory disease.**

4. The **vagina** is a tube that connects the uterus with the vestibule.

a. Superiorly, the vagina encircles the cervix of the uterus at the **fornix,** and forms the ectocervix which ends at the endocervical canal adjacent to the external os. 95% of cervical neoplasms develop at the transition zone of the ectocervix of the vagina and the endocervical canal.

b. The vagina passes through the urogenital hiatus of the pelvic diaphragm and the urogenital diaphragm.

c. Inferiorly, the external orifice of the vagina opens into the **vestibule** in the midline between the urethra and the perineal body.

d. The **posterior fornix** of the vagina is covered externally by peritoneum of the broad ligament, which is reflected onto the anterior aspect of the rectum and forms the **rectouterine pouch of Douglas.**

e. The vagina functions as the female organ of copulation and serves as a passageway for menstrual fluid and the fetus and placenta during labor and delivery.

EPISIOTOMY

An **episiotomy** is an incision of the posterior wall of the vagina, which may be performed during labor.

- A **median episiotomy** extends posteriorly in the midline through the perineal body.
- A **mediolateral episiotomy** extends through the bulbospongiosus and transversus perineus muscles. There is less risk of incising the fibers of the external anal sphincter in a mediolateral episiotomy.

5. The **broad ligament** is a fold of peritoneum that forms a mesentery suspending the uterine tubes, the ovaries, and most of the uterus.
   a. The **mesometrium** is the part of the broad ligament adjacent to the uterus, the **mesosalpinx** suspends the uterine tubes, and the **mesovarium** suspends the ovaries.
   b. Remnants of the **gubernaculum,** the round ligament of the uterus, and the proper ovarian ligament course in the mesometrium and mesovarium, respectively.
   c. The base of the broad ligament contains the **transverse cervical (cardinal) ligaments**.
      (1) The **transverse cervical ligaments** consist of condensations of connective tissue and support the cervix and the vagina.
      (2) The **uterine artery and vein** and the ureter course through the transverse cervical ligaments; adjacent to the cervix, the uterine artery crosses superior to the ureter (Figure 5–4).

HYSTERECTOMY PROCEDURE AND THE URETER

*In a **hysterectomy** procedure, a ureter may be injured or inadvertently ligated because of the proximity of the ureter to the cervix and to the uterine artery.*

IV. The **perineum** contains the urogenital diaphragm and the roots of the external genitalia.
   A. The **urogenital diaphragm** is situated inferior to the pelvic diaphragm and consists of skeletal muscles. It is enclosed by fascia (Figures 5–2 through 5–4).
      1. The urogenital diaphragm is traversed by the urethra in both sexes and by the vagina in females.
      2. The urogenital diaphragm contains the **deep transversus perineus muscles** and the **sphincter urethrae** (external urethral sphincter), which encircles the urethra.
      3. The **superior fascia** of the urogenital diaphragm covers the superior aspect of the urogenital diaphragm.
      4. The **inferior fascia** of the urogenital diaphragm or perineal membrane covers the inferior aspect of the urogenital diaphragm and is a site of attachment for the erectile bodies and skeletal muscles of the roots of the penis and clitoris.
   B. The **deep perineal pouch** contains the **sphincter urethrae** and the **deep transversus perineus** muscles in both sexes (see Figures 5–3 and 5–4).
      1. The deep perineal pouch is found between the superior and inferior fascial layers of the urogenital diaphragm.
      2. In males, the deep perineal pouch contains the **membranous part of the urethra** and the **bulbourethral (Cowper's) glands;** the ducts of the bulbourethral glands open into the penile urethra.
      3. In females, the deep perineal pouch contains the **urethra** and the **vagina.**
   C. The **superficial perineal pouch** contains the **crura and bulbs of the penis and clitoris,** which consist of erectile tissue.
      1. The 2 crura are attached to the ischiopubic rami and perineal membrane and continue as the corpora cavernosa into the penis and clitoris.
      2. In males, the bulb of the penis lies in the midline, is attached to the perineal membrane, and contains the penile urethra.

3. In females, the bulbs of the vestibule are situated in the wall of the vestibule with the greater vestibular (Bartholin's) glands.

4. Skeletal muscles cover the crura and the bulbs.

    a. The **ischiocavernosus muscles** cover the crura, and the **bulbospongiosus muscles** cover the bulbs (Figure 5–4). Both muscles contract during the sexual reflexes to limit venous drainage from the erectile tissues and help maintain erection.

    b. In males, the bulbospongiosus muscles contract and expel sperm and seminal fluid from the penile urethra during ejaculation and urine during micturition.

    c. The **superficial transversus perineus muscles** attach to the perineal body posterior to the bulb of the penis in males and the vestibule in females. These muscles contract to support the perineal body.

5. Colles' fascia forms the superficial fascial boundary of the superficial perineal space. It is continuous with dartos fascia of the scrotum, with superficial penile fascia that covers the penis and clitoris, and with Scarpa's fascia of the abdominal wall.

PENILE URETHRA LACERATION

Laceration of the penile urethra *may result in the extravasation of urine into the superficial perineal pouch that may spread into regions covered by fascia that is continuous with Colles' fascia. The extravasated urine may be found around the penis, in the scrotum, and deep to Scarpa's fascia on the anterior abdominal wall. The major difference between the male perineal pouches and the female perineal pouches is the location of the bulbourethral glands and the greater vestibular glands.*

- *The **bulbourethral glands** are situated in the deep pouch in the male.*
- *The **greater vestibular glands** are situated in the **superficial pouch** in the female.*

  D. The **external genitalia** in the male include the penis, the testes, and the scrotum (Figures 5–3 and 5–5). The penis consists of 2 corpora cavernosa and a corpus spongiosum.

    1. The **corpus spongiosum** is a continuation of the bulb of the penis and is situated on the ventral aspect of the penis. The corpus spongiosum surrounds the penile urethra and forms the glans penis.

    2. Paraurethral glands of Littré are adjacent to the penile urethra and function to lubricate the penile urethra.

    3. The **2 corpora cavernosa** are extensions of the crura and are situated on the dorsal aspect of the penis.

    4. Deep penile or Buck's fascia encloses the 3 erectile bodies, the deep dorsal vein, and the dorsal nerves and dorsal arteries of the penis.

  E. The **external genitalia** in the female include 2 pairs of labia, which surround the vestibule, and the clitoris (Figure 5–4).

    1. The **labia majora** contain connective tissue and fat and are the homologues of the scrotum; the round ligaments of the uterus end blindly in the labia majora.

    2. The **labia minora** lie medial to the labia majora and are in the lateral walls of the vestibule.

    3. The **vestibule** is a chamber into which the urethra and vagina open after passing through the urogenital diaphragm.

a. The **urethra** opens into the vestibule in the midline just posterior to the clitoris. Adjacent to the urethra are **paraurethral (Skene's) glands,** which are homologous to the prostate; their secretions lubricate the urethra and the vestibule.

    b. The **vagina** opens into the vestibule in the midline posterior to the urethra.

    c. The **bulbs of the vestibule** lie in the walls of the vestibule and join at the base of the clitoris.

   4. The **clitoris** consists of the 2 corpora cavernosa, which form both the body and the glans of the clitoris.

V. Direct branches of the aorta supply the rectum and the gonads.

   A. The **superior rectal artery,** which arises from the inferior mesenteric artery and with the middle rectal artery, supplies the rectum and anal canal above the pectinate line.

   B. The **testicular and ovarian arteries** arise between the renal arteries and the inferior mesenteric artery. The ovarian arteries cross the pelvic brim to supply the ovaries; the testicular arteries do not cross the pelvic brim but pass through the inguinal canals to supply the testes.

VI. Branches of the internal iliac artery supply pelvic viscera and perineal structures as well as the walls of the pelvis, the gluteal region, and the medial aspect of the thigh (Figure 5–9).

   A. **Pelvic Viscera Branches**

    1. The **umbilical artery** gives rise to 1 or 2 superior vesical arteries that supply the bladder. The umbilical artery is obliterated distal to the superior vesical arteries and forms a medial umbilical ligament.

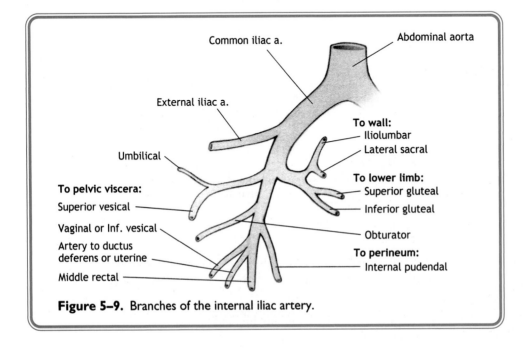

Figure 5–9. Branches of the internal iliac artery.

2. The **uterine artery** (females) is usually a branch of the umbilical artery that courses through the transverse cervical ligament and passes over the ureter to reach the uterine cervix. It courses through the broad ligament to supply the cervix, body, and fundus of the uterus and the uterine tubes.

3. The **artery to the ductus deferens** (males) usually arises from the inferior vesical artery and supplies the ductus deferens and the testis.

4. The **vaginal artery** (females) is usually a branch of the uterine artery that courses through the transverse cervical ligament and under the ureter to supply the bladder and the vagina above the pelvic diaphragm.

5. The **inferior vesical artery** (males) supplies the inferior aspect of the bladder and the prostate.

6. The middle rectal artery is usually a branch of the internal pudendal artery. It supplies the rectum and anal canal above the pectinate line and the seminal vesicles.

B. **Perineal Branches**

1. The **internal pudendal artery** supplies structures in the perineum.

2. It reaches the perineum by exiting the pelvis through the greater sciatic foramen, crossing the ischial spine, and passing through the lesser sciatic foramen.

3. The internal pudendal artery courses through the pudendal canal in the lateral wall of the ischioanal fossa.

   a. The **inferior rectal artery** supplies the anal canal below the pelvic diaphragm.

   b. The **perineal branch** supplies muscles in the superficial and deep perineal pouches and skin of the labia, vestibule, and scrotum.

   c. The **artery to the bulb** and a **deep artery of the penis** and clitoris supply erectile tissues, the bulbourethral gland, and the greater vestibular gland.

   d. The **dorsal artery of the penis** and clitoris supplies erectile tissue and skin of the penis and clitoris, respectively.

C. **Pelvic Wall Branches**

1. The **iliolumbar** artery crosses the pelvic brim to supply the iliacus, the psoas major and minor, and the quadratus lumborum muscles.

2. The **lateral sacral artery** supplies structures in the walls of the lesser pelvis and sends branches into the sacral canal through the ventral sacral foramina.

D. **Lower Limb Branches**

1. The **superior gluteal artery** exits the pelvis through the greater sciatic foramen and supplies the gluteus maximus, medius, and minimus muscles.

2. The **inferior gluteal artery** exits the pelvis through the greater sciatic foramen and supplies the gluteus maximus muscle.

3. The **obturator artery** exits the lesser pelvis with the obturator nerve through the obturator foramen and supplies the medial aspect of the thigh.

VII. The **pudendal nerve** innervates skeletal muscles in the pelvic diaphragm, in both perineal pouches of the perineum, and supplies the skin that overlies the perineum.

A. The pudendal nerve is formed by the **ventral rami of sacral spinal nerves** from S2, S3, and S4.

B. The pudendal nerve exits the pelvis superior to the pelvic diaphragm by passing through the greater sciatic foramen, crosses the ischial spine, and enters the pudendal canal after passing through the lesser sciatic foramen.

C. The pudendal nerve has 3 main branches.
   1. The **inferior rectal nerve** branches from the pudendal at the entrance of the pudendal canal, crosses the ischioanal fossa, and innervates the external sphincter muscle, skin of the anal canal below the pectinate line, and skin covering the anal triangle.
   2. The **perineal nerve** has superficial branches, the posterior scrotal nerves, that supply the skin of posterior scrotum and posterior labial nerves that supply the skin of the labia majora and vestibule. It also has a deep branch that innervates muscles in the superficial and deep perineal pouches, including both transverse perineus, the sphincter urethrae, bulbospongiosus, and ischiocavernosus muscles.
   3. The **dorsal nerve of the penis** and clitoris innervates skin of the penis and clitoris, respectively.

D. The pudendal nerve contains general sensory fibers, including those that convey pain from structures in the perineum (the lower part of the vagina, urethra, anal canal below the pectinate line, external genitalia, and skin of the perineum).

E. The pudendal nerve conveys sensations of pain during the second stage of labor that result from the stretching or tearing of the lower part of the vagina as the fetus passes through the pelvic and urogenital diaphragms.

PUDENDAL NERVE BLOCK

*A **pudendal nerve block** may be performed to suppress labor pain by anesthetizing the pudendal nerve as it crosses the **ischial spine**. The ischial spine may be palpated through the lateral wall of the vagina.*

F. The pudendal nerve has several **motor functions.**
   1. The pudendal nerve **maintains voluntary urinary continence** by facilitating contraction of the sphincter urethrae.
   2. It **maintains voluntary fecal continence** by facilitating contraction of the external anal sphincter.
   3. It **supports pelvic viscera** and helps prevent prolapse of pelvic visceral structures by facilitating contraction of skeletal muscles in the pelvic diaphragm and the urogenital diaphragm.
   4. The pudendal nerve helps **maintain erection** of the penis and clitoris during sexual reflexes by facilitating contraction of the ischiocavernosus and bulbospongiosus muscles.
   5. The nerve facilitates expulsion of **urine from the penile urethra** during micturition and sperm and seminal fluid during ejaculation by facilitating contraction of the bulbospongiosus muscles.

VIII. **Lower thoracic and lumbar splanchnic nerves** provide the sympathetic innervation to smooth muscle and glands in the pelvis and perineum (Figure 5–10).
   A. These nerves consist of **preganglionic sympathetic axons** that arise from the T11 through L2 segments of the cord.

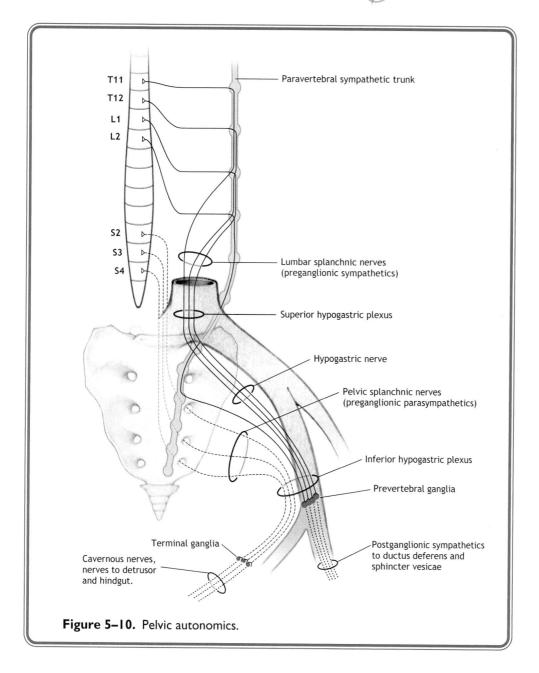

T11
T12
L1
L2

S2
S3
S4

Paravertebral sympathetic trunk

Lumbar splanchnic nerves
(preganglionic sympathetics)

Superior hypogastric plexus

Hypogastric nerve

Pelvic splanchnic nerves
(preganglionic parasympathetics)

Inferior hypogastric plexus

Prevertebral ganglia

Postganglionic sympathetics
to ductus deferens and
sphincter vesicae

Terminal ganglia

Cavernous nerves,
nerves to detrusor
and hindgut.

Figure 5–10. Pelvic autonomics.

B. Preganglionic sympathetic axons in these nerves either synapse in the **inferior mesenteric ganglion** or cross the pelvic brim in the **superior hypogastric plexus** to synapse in prevertebral ganglia in the **inferior hypogastric plexus** or in one of its subsidiary plexuses.

   1. The **superior hypogastric plexus** consists of sympathetic axons and visceral pain fibers.

   2. The visceral pain fibers course back into the spinal cord with lower thoracic and lumbar splanchnic nerves.

C. Lower thoracic and lumbar splanchnic nerves have several motor functions.

   1. They facilitate **contraction of the sphincter vesicae,** the smooth muscle at the base of the bladder, which permits the bladder to fill and prevents retrograde ejaculation.

   2. They facilitate **contraction of smooth muscle in the ductus deferens, prostate, and seminal vesicles** during emission, allowing sperm and the components of seminal fluid to reach the prostatic urethra.

   3. They facilitate **emptying of the secretory products of the bulbourethral and greater vestibular glands,** which lubricate the penile urethra and the vestibule, respectively.

   4. They facilitate **contraction of the internal anal sphincter,** which helps maintain fecal continence.

D. **Lower thoracic and lumbar splanchnic nerves** carry visceral pain sensations from gastrointestinal structures in the hindgut and from pelvic viscera, covered by peritoneum including the bladder and most of the uterus.

E. **Lower thoracic and lumbar splanchnic nerves** carry visceral pain from the fundus, body, and cervix of the uterus and the upper part of the vagina during the first stage of labor. Pain sensations from pelvic viscera may be referred over the T11 through L1 dermatomes.

IX. **Pelvic splanchnic nerves** (or pelvic nerves) provide the parasympathetic innervation to smooth muscle and glands in the pelvis and perineum (see Figure 5–8).

A. Pelvic splanchnic nerves contain **preganglionic parasympathetic axons** that arise from neurons in the S2, S3, and S4 spinal cord segments.

B. These nerves branch from the ventral rami of the S2, 3, and 4 spinal nerves as the rami emerge from the ventral sacral foramina.

C. Pelvic splanchnic nerves join with the sympathetic splanchnic nerves in the superior hypogastric plexus to form the **inferior hypogastric plexus**. The inferior hypogastric plexus contains both parasympathetic and sympathetic axons and sensory fibers.

D. These nerves synapse in **terminal ganglia** situated in the wall of the bladder and in the hindgut (distal third of the transverse colon to the pectinate line of the anal canal) and in the inferior hypogastric plexus.

E. Pelvic splanchnic nerves **contain sensory fibers** that respond to distension of the bladder, rectum, uterus, and vagina. Pelvic nerves convey pain from pelvic viscera not covered by peritoneum.

F. Pelvic splanchnic nerves have **several motor functions.**

1. Pelvic splanchnic nerves facilitate **bladder** emptying during micturition by facilitating contraction of the detrusor muscle in the bladder wall.
2. They **stimulate glandular secretions** and peristalsis in the hindgut.
3. They **facilitate emptying of the rectum** in defecation by relaxing the internal anal sphincter and facilitating contraction of the smooth muscle in the ampulla of the rectum.
4. They **facilitate erection** by causing the dilation of helicine arteries in erectile tissues.
5. Pelvic splanchnic nerves **stimulate the secretory activity** of the prostate, seminal vesicles, and bulbourethral and greater vestibular glands.

X. **Micturition** is the process by which urine flows from the bladder into the urethra. It is mediated by the vesical reflex.

  A. The vesical reflex has 2 components—a **filling stage** and an **emptying stage.**

    1. The filling stage

      a. During the **filling stage,** stretch receptors in the wall of the bladder that are innervated by sensory fibers in **pelvic splanchnic nerves** respond to the presence of urine in the bladder and the degree of stretch of the bladder wall.

      b. A **bladder storage center** is activated in the central nervous system (CNS), which stimulates sympathetic nerves to inhibit the detrusor, allowing the bladder wall to continue to stretch without contraction.

      c. As the bladder fills, the wall of the bladder compresses the oblique entry points of the ureters, **preventing reflux** of urine into the ureters.

      d. At or near a micturition threshold, conscious **perception of bladder fullness** is appreciated.

      e. **Pain fibers** that are stimulated by bladder fullness course with **thoracic and lower lumbar splanchnic nerves** and refer pain over the T11 through L1 dermatomes in the lower anterior abdominal wall.

    2. The emptying stage

      a. During the **emptying stage,** the **micturition threshold is reached,** and a CNS micturition center is activated.

      b. The **sphincter urethrae and sphincter vesicae are relaxed,** pelvic splanchnic nerves facilitate contraction of the detrusor, and urine passes through the length of the urethra.

      c. The **bulbospongiosus muscle** in the male contracts to expel urine from the penile urethra near the end of micturition.

      d. The **thoracoabdominal diaphragm** and the abdominal muscles contract to increase intrapelvic pressure on the bladder, accelerating the flow of urine.

  B. The vesical reflex is facilitated by parasympathetic and sympathetic nerves and by the pudendal nerves.

DISORDERS OF MICTURITION

*Lesions that interrupt the **CNS control centers** for storage or micturition or the cortical pathways that control the CNS micturition centers cause **supraspinal bladder disorders.***

The vesical reflex is intact because the sensory and motor fibers in the parasympathetic, sympathetic, and pudendal nerves are unaffected.

*An **uninhibited or infantile bladder** may result from lesions to cortical areas that provide control over the micturition storage and emptying centers. In these patients, the bladder fills normally but empties suddenly and completely, as in an infant, with little risk of infection from the presence of residual urine.*

*A **spinal, automatic, or spastic bladder** may result from lesions to the spinal cord above sacral levels. In these patients, parasympathetic neurons that innervate the detrusor muscle are not inhibited effectively when the bladder is stretched during filling. The detrusor contracts in response to a minimum amount of stretch, causing frequent emptying.*

The bladder tends to be small and contracted. Residual urine may be present after emptying, increasing the risk of infection.

*An **atonic bladder** may result from lesions to the sacral spinal cord or to the roots of sacral spinal nerves in the cauda equina. These lesions disrupt the neural components of the vesical reflex.*

In these patients, the bladder fills to capacity, but urine dribbles through the urethra continuously because the detrusor fails to contract and empty the bladder, and the voluntary urethral sphincter may be weakened.

Patients with an atonic bladder tend to retain a considerable volume of urine with a high infection risk and pass urine only as a result of overflow incontinence.

    XI. Defecation is the process by which the contents of the rectum are emptied.

        A. Visceral afferent fibers mediating distension of the rectum and anal canal above the pectinate line course back into the sacral spinal cord in pelvic splanchnic nerves.

        B. Pelvic splanchnic nerves facilitate peristalsis by stimulating contractions of smooth muscle in the ampulla of the rectum.

        C. The **internal anal sphincter and the external anal sphincter** are inhibited, which facilitates an emptying of rectal contents.

    XII. The male and female sexual reflexes include erection, secretion, emission, and ejaculation. These reflexes use parasympathetic and sympathetic nerves and the pudendal nerve.

        A. Erection is facilitated mainly by pelvic splanchnic nerves and by postganglionic cavernous nerves.

            1. Erection results from a relaxation of the vascular smooth muscle, and dilation, uncoiling, and filling of the arteries in the erectile tissues of the crura and corpora cavernosa, the bulb, and the corpus spongiosum.

            2. Contractions of the ischiocavernosus and bulbospongiosus muscles (innervated by perineal branches of the pudendal nerve) help maintain erection.

INJURY TO CAVERNOUS NERVES

In surgical procedures involving the prostate, the cavernous nerves may be lesioned.

- *The cavernous nerves course lateral to the prostate before passing through the urogenital hiatus to enter the perineum.*
- *In these patients, **impotence** (an inability to obtain an erection) may result.*

CLINICAL CORRELATION

        B. Secretion is also facilitated by pelvic splanchnic nerves.

            1. In the male, the **pelvic splanchnic nerves** stimulate the secretory activity of the prostate and the seminal vesicles and the secretory activity of bulbourethral glands.

2. In the female, the **pelvic splanchnic nerves** stimulate the secretory activity of vaginal glands and the greater vestibular glands; the vagina becomes elongated, and the uterus moves from an anteverted or a retroverted position to a midposition in the pelvis.

C. **Emission** is facilitated by **lower thoracic and lumbar splanchnic nerves.**
   1. During emission, contraction of the smooth muscle in the ductus deferens and smooth muscle in the prostate and seminal vesicles in males promotes movement of sperm and seminal fluid through the ductus deferens and ejaculatory duct and into the prostatic urethra. Secretions from the prostate also enter the prostatic urethra.
   2. Contraction of the sphincter vesicae muscle at the base of the bladder prevents **retrograde ejaculation,** or the reflux of seminal fluid into the bladder.

D. **Ejaculation** closely follows emission and is facilitated by the pudendal nerve.
   1. During ejaculation, the **sphincter urethrae** is relaxed.
   2. During ejaculation, rhythmic contractions of the **bulbospongiosus muscle** propel sperm and seminal fluid through the membranous and penile urethra.

CLINICAL PROBLEMS

Match the clinical feature in Questions 1–10 with the appropriate site of a lesion to nerve or nerves in choices A–E. (A choice may be used once, more than once, or not at all.)

   A. Pelvic splanchnic nerves
   B. Lumbar splanchnic nerves
   C. Inferior rectal nerve
   D. Perineal nerve
   E. Dorsal nerve of the penis

1. Weakness of the internal anal sphincter

2. Weakness in the ability to contract the bladder

3. Impotence (inability to obtain an erection)

4. Inability to sense bladder fullness

5. Weakness of the sphincter urethrae

6. Sperm and seminal fluid enter the bladder during ejaculation

7. Weakness of external anal sphincter

8. Pain from external hemorrhoids

9. Difficulty maintaining erection

10. Weakness of urogenital diaphragm

An 18-year-old male lacerates the penile urethra in the bulb of the penis. Urine has extravasated from the urethra into areas that are covered by a continuation of the fascia over the bulb.

11. Where might the extravasated urine be found?
    A. In the ischioanal fossa
    B. In the medial thigh
    C. In the anterior abdominal wall
    D. In the rectus sheath deep to the rectus abdominis muscle
    E. In the gluteal region

A pregnant female undergoes a pudendal nerve block to reduce pain during labor and delivery.

12. What might the patient experience as a result of this procedure?
    A. An inability to contract and empty the bladder
    B. An inability to perceive bladder fullness
    C. An inability to empty the rectum
    D. Weakness of the voluntary sphincter of the bladder
    E. Inability to contract muscles of the anterior abdominal wall

13. What is the location of the anesthetized muscle?
    A. Superficial perineal pouch
    B. Ischioanal fossa
    C. Lesser pelvis
    D. Above the pelvic diaphragm
    E. Urogenital diaphragm

In the weeks after delivery, a new mother develops a prolapsed uterus.

14. What structure most likely was weakened during delivery and resulted in this condition?
    A. Piriformis muscle
    B. Suspensory ligament
    C. Sacrospinous ligament
    D. Pubococcygeus muscle
    E. Superficial transversus perinei muscle

A 45-year-old man has problems with defecation, frequent bladder infections, and difficulty engaging in sexual relations. All of his problems are caused by spondylolisthesis of the sacrum, which compresses sacral roots in the cauda equina.

15. Which of the following might describe 1 of the patient's difficulties?
    A. The bladder fills normally but empties suddenly and completely.
    B. The bladder contracts in response to a minimum amount of stretch, causing frequent involuntary reflex emptying.

C. The bladder fills to capacity, but urine dribbles through the urethra continuously because the detrusor fails to contract and empty the bladder.

D. Seminal fluid accumulates in the bladder during emission and ejaculation.

E. There is an absence of peristalsis in the ductus deferens.

During prostatic surgery, nerves that are secretomotor to the prostate are lesioned.

16. What signs or symptoms might the patient experience?

A. Impotence

B. Inability to contract the levator ani muscle

C. Altered sensation in skin of the scrotum

D. Absence of peristalsis in the ductus deferens

E. Retrograde ejaculation of seminal fluid into the bladder

A mass has compressed a common iliac artery.

17. To which of the following structures will arterial blood supply be unaffected?

A. Detrusor muscle

B. Gluteus maximus muscle

C. Testis

D. Rectus abdominis muscle

E. Corpus spongiosum

Nerves containing axons of motor neuron cell bodies found in the L2 segment of the spinal cord are lesioned.

18. What signs or symptoms might the patient experience?

A. Weakness in the ability to empty the bladder

B. Absence of peristalsis in the rectum

C. Weakness of the hamstring muscles

D. Weakness of the internal urethral sphincter

E. Fecal incontinence

A young female patient is driven to the emergency room and, upon examination, appears to be in a state of shock. She shows signs of an internal hemorrhage. A vaginal exam reveals that the patient's cervix is soft. The patient indicates that she missed her last menstrual period and has a history of pelvic inflammatory disease. Your diagnosis is that of a ruptured ectopic pregnancy.

19. Which is the most likely site of the ectopic implantation?

A. Broad ligament

B. Body of the uterus

C. Ampulla of the uterine tube

D. Posterior fornix of the vagina

E. Infundibulum of the uterine tube

20. Your examination of the patient in Question 19 reveals a palpable fluid-like mass. The mass probably contains the contents of the ruptured implantation. The mass is most likely in the:

    A. Omental bursa

    B. Rectouterine pouch

    C. Vesicouterine pouch

    D. Posterior fornix of the vagina

    E. Broad ligament

ANSWERS

1. A

2. A

3. A

4. A

5. D

6. B

7. C

8. C

9. D

10. D

11. The answer is C. Laceration of the penile urethra may result in the extravasation of urine into the superficial perineal pouch and on to the anterior abdominal wall.

12. The answer is D. The sphincter urethrae muscle, the only choice innervated by the pudendal, which contracts to maintain urinary continence, will be weakened.

13. The answer is E. The sphincter urethrae is found in the urogenital diaphragm.

14. The answer is D. The pelvic diaphragm, in particular the pubococcygeus muscle, helps prevent prolapse or herniation of pelvic viscera after delivery.

15. The answer is C. The patient has an atonic bladder; these patients tend to retain a considerable volume of urine with a high infection risk and pass urine only as a result of overflow incontinence.

16. The answer is A. In surgical procedures involving the prostate, the cavernous nerves may be lesioned. In these patients, impotence, or an inability to obtain an erection, may result.

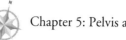

17. The answer is C. The testis is supplied by a testicular artery, which arises from the abdominal aorta proximal to the site of the blockage. All other choices are supplied by branches of the internal iliac artery distal to the blockage.

18. The answer is D. The internal urethral sphincter is innervated by sympathetic axons, which course in lower thoracic and lumbar splanchnic nerves, including L2. All other choices are innervated by spinal cord segments below L2.

19. The answer is C. The most common site of an ectopic implantation is in the ampulla of the uterine tube.

20. The answer is B. In the upright position, fluid in the peritoneal cavity tends to accumulate in the rectouterine pouch of Douglas.

CHAPTER 6
UPPER LIMB

I. Bones and Joints

A. The **pectoral girdle** consists of the **clavicle** and **scapula** and suspends the humerus and the upper limb away from the axial skeleton to increase range of movement (Figure 6–1).

   1. The medial end of the clavicle articulates with the sternum at the **sternoclavicular joint**, the only articulation between the axial skeleton and the upper limb.

      a. The sternoclavicular joint contains an **articular disk** that divides the joint into 2 joint cavities.

      b. At the joint between the articular disk and the sternum, the clavicle can be protracted and retracted.

      c. At the joint between the articular disk and the medial end of the clavicle, the clavicle can be elevated and depressed.

      d. The clavicle can be rotated about its long axis at the sternoclavicular joint.

   2. The lateral end of the clavicle articulates with the acromion process of the scapula at the **acromioclavicular joint**.

FRACTURE OF THE CLAVICLE

*The **clavicle is commonly fractured** at its weakest point between the middle third and the lateral third.*
* *The medial two-thirds of the clavicle may be elevated by the **sternocleidomastoid muscle**, and the lateral third may be depressed by the weight of the limb or **adducted by the pectoralis major**.*
* *The ventral rami of C8 and T1 in the **medial cord** of the brachial plexus may be lacerated as a result of the fracture.*

SHOULDER TRAUMA TO THE ACROMIOCLAVICULAR JOINT

***Shoulder trauma** may cause a **subluxation** of the acromion at the acromioclavicular joint. The **coracoclavicular ligament**, which extends from the acromion to the coracoid process of the scapula, prevents dislocation at the acromioclavicular joint.*

   3. The **pectoral girdle** can be elevated, depressed, protracted, and retracted on the chest wall (Table 6–1).

B. The **glenohumeral joint** is a **ball-and-socket** joint. It permits flexion and extension, abduction and adduction, and medial and lateral rotation (Table 6–2).

   1. At this joint, the **head of the humerus** articulates with the **glenoid fossa of the scapula**.

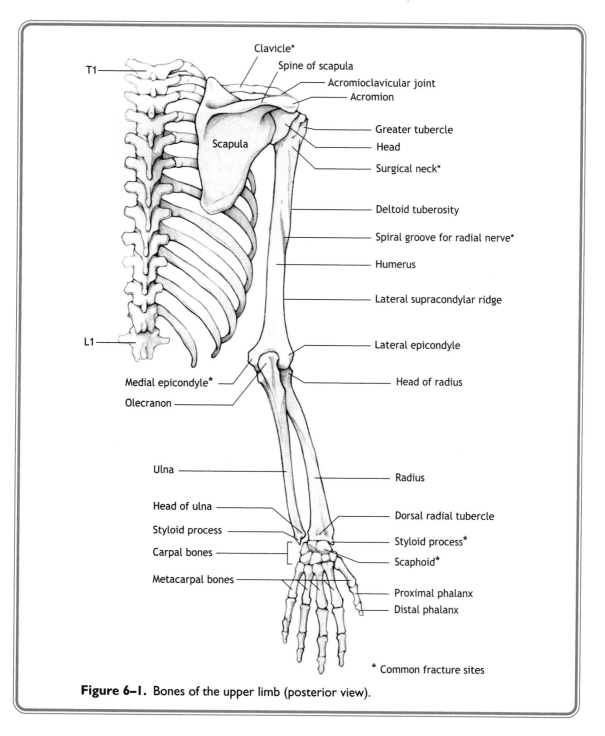

Figure 6–1. Bones of the upper limb (posterior view).

Table 6–1. Muscles that act at the pectoral (shoulder) girdle.

Action	Muscles Involved	Innervation	Major Segments of Innervation
Elevation	Levator scapulae Trapezius, upper part	Dorsal scapular Accessory	C4, 5
Depression	Pectoralis minor Trapezius, lower part Latissimus dorsi	Medial pectoral Accessory Thoracodorsal	C7, 8 C6, 7, 8
Protraction	Serratus anterior Pectoralis minor	Long thoracic Medial pectoral	C5, 6, 7 C7, 8
Retraction	Rhomboid major and minor Trapezius, middle fibers	Dorsal scapular Accessory	C5 C1-5
Lateral rotation of scapula (in abduction)	Serratus anterior, lower half Trapezius, upper and lower parts	Long thoracic Accessory	C5, 6, 7 C1-5
Medial rotation of scapula (in adduction)	Rhomboid major and minor Levator scapulae	Dorsal scapular Dorsal scapular	C5 C4, 5

2. The scapula is laterally (upwardly) or medially (downwardly) rotated along the chest wall during abduction and adduction of the humerus at the glenohumeral joint.
   a. In abduction, for every 2 degrees of abduction of the arm at the glenohumeral joint, there is 1 degree of lateral or upward rotation of the scapula.
   b. In 180 degrees of full abduction, there is approximately 120 degrees of abduction of the humerus at the glenohumeral joint, and approximately 60 degrees of scapular rotation.
3. The **tendons of rotator cuff** muscles strengthen the articular capsule of the glenohumeral joint and include the **supraspinatus, infraspinatus, teres minor, and subscapularis** (the SITS muscles).

INFLAMMATION OF THE ROTATOR CUFF

The **tendons of muscles of the rotator cuff** may become torn or inflamed.
• The tendon of the supraspinatus is most commonly affected.

Patients with rotator cuff tears experience pain anterior and superior to the glenohumeral joint during abduction.

4. The **capsule of the glenohumeral joint** is strengthened by several ligaments.
   a. The **glenohumeral bands** strengthen the anterior aspect of the joint.
   b. The **coracohumeral ligament** strengthens the superior aspect of the joint.
5. The **coracoacromial ligament** prevents superior displacement of the head of the humerus.

Table 6–2. Muscles that act at the shoulder (glenohumeral) joint.

Action	Muscles Involved	Innervation	Major Segments of Innervation
Flexion of humerus	Pectoralis minor, clavicular head	Lateral Pectoral	C5, 6, 7
	Deltoid, clavicular part	Axillary	C5, 6
	Biceps, short head	Musculocutaneous	C5, 6
	Coracobrachialis	Musculocutaneous	C6, 7
Extension of humerus	Deltoid, posterior fibers	Axillary	C5, 6
	Latissimus dorsi	Thoracodorsal	C6, 7, 8
	Teres major	Lower Subscapular	C6
Abduction of humerus	Deltoid, middle fibers	Axillary	C5, 6
	Supraspinatus	Suprascapular	C5
Adduction of humerus	Pectoralis major, sternocostal part	Medial and Lateral Pectoral	C6-T1
	Latissimus dorsi	Thoracodorsal	C6, 7, 8
	Teres major	Lower Subscapular	C5, 6
Lateral rotation of scapula in abduction	Deltoid, posterior fibers	Axillary	C5, 6
	Infraspinatus	Suprascapular	C5, 6
	Teres minor	Axillary	C6
Medial rotation of scapula in adduction	Pectoralis major	Medial and lateral Pectoral	C5-T1
	Latissimus dorsi	Thoracodorsal	C6, 7, 8
	Deltoid, clavicular part	Axillary	C5, 6, 7
	Teres major	Lower Subscapular	C5, 6
	Subscapularis	Upper and Lower Subscapular	C5, 6

HUMERAL DISLOCATION

In a **dislocation of the humerus at the glenohumeral joint**, the **head of the humerus** is commonly displaced inferiorly and then anteriorly and becomes positioned just inferior to the **coracoid process**. A dislocation of the head of the humerus may stretch the **axillary nerve** or the **radial nerve**.

HUMERAL FRACTURE

In a **fracture of the surgical neck of the humerus**, the axillary nerve may be lesioned, and the posterior circumflex humeral artery may be lacerated.

A fracture of the **greater tubercle** of the humerus may result in avulsion of the greater tubercle and detachment of the rotator cuff muscles from the humerus. In patients with fractures of the greater tubercle, the remaining rotator cuff muscle, the subscapularis, medially rotates the humerus at the glenohumeral joint.

A **transverse fracture** of the humerus distal to the deltoid tuberosity may result in abduction of the proximal fragment by the deltoid muscle.

*In a **midshaft (spiral) fracture** of the humerus, the radial nerve may be lesioned, and the profunda brachial artery may be lacerated.*

*In patients with a **supracondylar fracture of the humerus**, contractions of the triceps and the brachialis may shorten the arm. The **median nerve** may be lesioned as a result of an intercondylar or supracondylar fracture of the distal end of the humerus.*

    C. The **elbow joints** include the **humeroradial joint,** the **humeroulnar joint,** and the **proximal radioulnar joint.**

        1. At the **humeroradial joint**, the head of the radius articulates with the capitulum of the humerus (see Figure 6–1).

        2. At the **humeroulnar joint**, the trochlear notch of the ulna articulates with the trochlea of the humerus.

        3. The **humeroradial joint** and the **humeroulnar joint** are hinge joints that permit flexion and extension (Table 6–3).

        4. At the **proximal radioulnar joint** the radial notch of the ulna articulates with the head of the radius; pronation and supination occur at this joint and at the distal radioulnar joint (Table 6–4).

EPICONDYLITIS

Lateral epicondylitis (tennis elbow) is an inflammation of the common extensor tendon that results from forced extension and flexion of the forearm at the elbow. Patients exhibit pain over the lateral epicondyle, which may radiate down the posterior aspect of the forearm.

Medial epicondylitis (golfer's elbow) is an inflammation of the common flexor tendon that results from repetitive flexion and pronation of the forearm at the elbow.

FRACTURE OF THE MEDIAL EPICONDYLE

*In a **fracture of the medial epicondyle** of the humerus, the ulnar nerve may be lesioned.*

    D. At the **wrist**, the **carpal bones** articulate proximally with the radius and the ulna at the **radiocarpal** and **ulnocarpal** joints and distally with the metacarpals at carpometacarpal joints (see Figure 6–1).

        1. The radiocarpal joint is formed by the distal end of the radius and the scaphoid and the lunate; the ulnocarpal joint is formed by the distal end of the ulna, an articular disk, and the triquetrum. The radiocarpal and ulnocarpal joints permit flexion, extension, abduction (radial deviation), and adduction (ulnar deviation) (Table 6–5).

Table 6–3. Muscles that act at the humeroulnar and humeroradial joints.

Action	Muscles Involved	Innervation	Major Segments of Innervation
Flexion of ulna and radius	Brachialis	Musculocutaneous	C5, 6
	Biceps brachii	Musculocutaneous	C5, 6
	Brachioradialis	Radial	C5, 6
Extension of ulna and radius	Triceps brachii	Radial	C7, 8

Table 6–4. Muscles that act at the proximal and distal radioulnar joints.

Action	Muscles Involved	Innervation	Major Segments of Innervation
Pronation (radius rotates over ulna)	Pronator teres Pronator quadratus	Median Median (ant. interosseous n.)	C6, 7 C8, T1
Supination (radius returns to anatomic position)	Supinator Biceps brachii	Radial (deep br.) Musculocutaneous	C6, 7, 8 C5, 6

COLLES' FRACTURE

*A **Colles' fracture** of the **distal radius** may result in **avulsion of the styloid process** from the shaft of the radius.*

- *The radius may be shortened, and the styloid process of the ulna may project further distally than the styloid process of the radius.*
- *In patients with a fracture of the distal radius, the forearm and hand may exhibit a **"dinner fork" deformity** as a result of the posterior displacement of the distal part of the radius.*

2. The **8 carpal bones** are loosely arranged into 2 rows.
   a. The **proximal row** contains, from lateral to medial, the **scaphoid, lunate, triquetrum, and pisiform** bones.
      (1) The pisiform is anterior to the triquetrum.
      (2) The pisiform is a sesamoid bone that is embedded in the tendon of the flexor carpi ulnaris.
   b. The distal row contains, from lateral to medial, the **trapezium, trapezoid, capitate, and hamate** bones.

Table 6–5. Muscles that act at the wrist joints.

Action	Muscles Involved	Innervation	Major Segments of Innervation
Flexion of hand	Flexor carpi ulnaris Flexor carpi radialis	Ulnar Median	C8 C6, 7
Extension of hand	Extensor carpi ulnaris Extensor carpi radialis longus Extensor carpi radialis brevis	Radial (deep br.) Radial Radial (deep br.)	C7, 8 C6, 7 C6, 7
Abduction (radial deviation) of hand	Extensor carpi radialis longus/brevis Flexor carpi radialis	Radial (deep br.) Median	C6, 7 C6, 7
Adduction (ulnar deviation) of hand	Flexor carpi ulnaris Extensor carpi ulnaris	Ulnar Radial (deep br.)	C8 C7, 8

LUNATE DISLOCATION

*The **lunate** is the most commonly dislocated carpal bone.*
- *The lunate is typically dislocated anteriorly into the **carpal tunnel**.*
- *Dislocation of the lunate may cause carpal tunnel syndrome (see later discussion).*

> **3.** The carpal tunnel is formed posteriorly by the 8 carpal bones.
>   **a.** The **flexor retinaculum** completes the carpal tunnel anteriorly; it attaches medially to the pisiform and hamate and laterally to the tubercles of the scaphoid and trapezium.
>   **b.** The **carpal tunnel** contains the **median nerve and 9 tendons** that arise from 3 flexor muscles—the flexor digitorum superficialis and profundus and the flexor pollicis longus—in the anterior forearm.
> **4.** The **canal of Guyon** is situated between the pisiform and the hook of the hamate superficial to the carpal tunnel. The ulnar nerve, ulnar artery, and ulnar vein cross the wrist and pass into the hand after traversing the canal of Guyon.

E. In each of the 4 fingers, **a metacarpal** and **the 3 phalanges form 3 joints:** a **metacarpophalangeal** (MP) joint, a **proximal interphalangeal** (PIP) joint, and a **distal interphalangeal** (DIP) joint.
  1. At the **MP joints**, the metacarpals articulate with the proximal phalanges. The MP joints are condyloid joints and permit flexion, extension, abduction, and adduction (Table 6–6).
  2. At the **PIP joints**, the proximal phalanges articulate with the middle phalanges. At **the DIP joints**, the middle phalanges articulate with the distal phalanges. The PIP and DIP joints are hinge joints that permit flexion and extension (see Table 6–6).

F. The **thumb** has 3 joints: a **carpometacarpal** joint, an **MP** joint, and an **interphalangeal** joint.
  1. The carpometacarpal joint is a saddle joint formed by the first metacarpal and by the trapezium. It permits flexion, extension, abduction, adduction, and rotation (Table 6–7).
  2. Actions at the MP joint and the interphalangeal joint are similar to those at these joints in the fingers.

G. The tendons of the 2 extensor muscles of the thumb and the abductor pollicis longus form the boundaries of the **"anatomic snuffbox,"** a small region of the posterolateral part of the wrist.
  1. The **extensor pollicis brevis** and the **abductor pollicis longus** form the lateral border of the anatomic snuffbox.
  2. The **extensor pollicis longus** forms the medial border of the anatomic snuffbox.
    a. The **scaphoid** and **trapezium** bones are in the floor of the snuffbox.
    b. The **radial artery** passes through the snuffbox, and the superficial branch of the radial nerve innervates skin over the snuffbox.

SCAPHOID FRACTURE

*The **scaphoid** is the most commonly fractured carpal bone.*
- *Patients with a scaphoid fracture may exhibit pain and tenderness localized over the **anatomic snuffbox**. The proximal part of the scaphoid may undergo avascular necrosis because the blood supply to the bone supplies the distal part first and then the proximal part.*

Table 6–6. Muscles that act at the joints of the fingers.

Action	Muscles Involved	Innervation	Major Segments of Innervation
Flexion, all fingers			
at MP joints	Lumbricals	Median, index and middle	C8, T1
		Ulnar (deep br.), ring and little	C8, T1
	Interossei	Ulnar (deep br.)	C8, T1
at MP joint of little finger	Flexor digiti minimi	Ulnar (deep br.)	C8, T1
at PIP joints	Flexor digitorum superficialis	Median	C7, C8, T1
	Lumbricals	Median, index and middle	C8, T1
		Ulnar (deep br.), ring and little	C8, T1
at DIP joints	Flexor digitorum profundus	To index and middle Median (ant. interosseous n.) to ring and little Ulnar (deep br.)	C8
Extension			
at MP joints of all fingers	Extensor digitorum	Radial (deep br.)	C6, 7, 8
of index finger	Extensor indicis	Radial (deep br.)	C7, 8
of little finger	Extensor digiti minimi	Radial (deep br.)	C7, 8
Extension			
at IP joints	Lumbricals	Median, index and middle	C8, T1
		Ulnar (deep br.), ring and little	C8, T1
	Interossei	Ulnar (deep br.)	C8, T1
Abduction			
at MP joints	Dorsal interossei	Ulnar (deep br.)	C8, T1
	Abductor digiti minimi	Ulnar (deep br.)	C8, T1

continued

Table 6–6. Muscles that act at the joints of the fingers. (*continued*)

Action	Muscles Involved	Innervation	Major Segments of Innervation
Adduction			
at MP joints of all fingers except middle finger	Palmar interossei	Ulnar (deep br.)	C8, T1
Opposition			
of little finger	Opponens digiti minimi	Ulnar (deep br.)	C8, T1

MP, metacarpophalangeal joints; PIP, proximal interphalangeal joints; DIP distal interphalangea joints.

II. Arterial Blood Supply (Figure 6–2)

A. The **subclavian artery** arises from the brachiocephalic artery on the right and the arch of the aorta on the left and becomes the **axillary artery** at the lateral border of the first rib. The **thyrocervical trunk** of the subclavian gives rise to the **supras-capular** and **transverse cervical** arteries, which supply muscles that attach to the scapula.

Table 6–7. Muscles that act at joints of the thumb.

Action	Muscles Involved	Innervation	Major Segments of Innervation
Flexion			
at all joints, especially IP joint	Flexor pollicis longus	Median (ant. interosseous n.)	C8, T1
at MP joint	Flexor pollicis brevis	Median (recurrent br.)	C8, T1
Extension			
at all joints	Extensor pollicis longus	Radial (deep br.)	C7, 8
at MP joint	Extensor pollicis brevis	Radial (deep br.)	C7, 8
Abduction	Abductor pollicis longus	Radial (deep br.)	C7, 8
	Abductor pollicis brevis	Median (recurrent br.)	C8, T1
Adduction	Adductor pollicis	Ulnar (deep br.)	C8, T1
Opposition	Opponens pollicis	Median (recurrent br.)	C8, T1

IP, interphalangeal joints; MP, metacarpophalangeal joints.

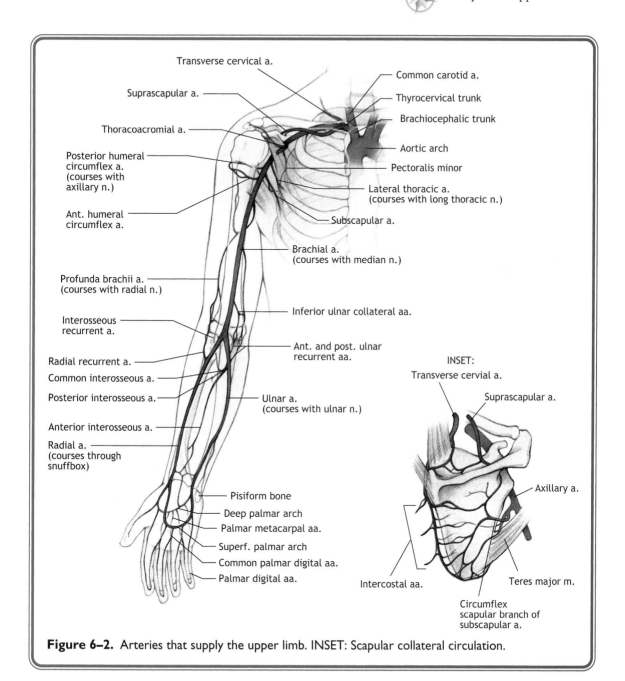

Figure 6–2. Arteries that supply the upper limb. INSET: Scapular collateral circulation.

1. The **suprascapular artery:**
    a. Crosses the posterior triangle of the neck.
    b. Passes over the transverse scapular ligament.
    c. Supplies the supraspinatus and infraspinatus on the posterior aspect of the scapula.
2. The **transverse cervical artery:**
    a. Crosses the posterior triangle of the neck.

b. Divides into a superficial branch that supplies the trapezius.

    c. Divides into a deep branch (the dorsal scapular artery) that supplies the rhomboids and the levator scapulae.

B. The **axillary artery** is divided into 3 parts by the pectoralis minor and has 6 branches: 1 from the first part, 2 from the second part, and 3 from the third part.

    1. The **superior thoracic artery** supplies the first 2 intercostal spaces and the serratus anterior.

    2. The **thoracoacromial artery** supplies the anterior wall of the axilla, including the pectoral muscles, deltoid, clavicle, and acromioclavicular joint.

    3. The **lateral thoracic artery** courses with the long thoracic nerve and supplies the serratus anterior, the pectoral muscles, and the breast.

    4. The **subscapular artery** is the largest branch of the axillary artery.

        a. It gives rise to the circumflex scapular artery, which courses around the lateral border of the scapula and supplies the teres major, teres minor, and infraspinatus muscles.

        b. It gives rise to a thoracodorsal artery, which supplies the latissimus dorsi muscle.

    5. The **posterior circumflex humeral artery** arises adjacent to the subscapular artery.

        a. It courses with the axillary nerve through the quadrangular space around the surgical neck of the humerus.

        b. It supplies the deltoid, teres major, and teres minor muscles and the long head of the triceps muscle.

    6. The **anterior circumflex humeral artery** is smaller than the posterior circumflex humeral and supplies the muscles in the anterior arm.

AXILLARY ARTERY OCCLUSION

*In an **occlusion of the first or second part of the axillary artery** or of the subclavian artery, the **circumflex scapular** and **thoracodorsal branches** of the subscapular artery contribute to collateral circulation, which may bypass the blockage (see INSET Figure 6–2). Anastomoses may develop superior and posterior to the scapula between the thoracodorsal and circumflex scapular branches of the subscapular artery and the suprascapular, dorsal scapular, and posterior intercostal arteries.*

C. The **brachial artery** continues from the axillary at the lower border of the teres major tendon and supplies the anterior and posterior aspects of the arm.

    1. The **profunda brachial artery** spirals around the midshaft of the humerus with the radial nerve.

    2. The profunda brachial artery supplies the triceps brachii muscle and contributes to collateral circulation around the elbow.

VOLKMANN'S ISCHEMIC CONTRACTURE

Volkmann's ischemic contracture *may be caused by a supracondylar fracture of the humerus. Displacement of the humerus as a result of the fracture may compress the **brachial artery** and result in ischemia of the forearm and hand. In these patients, the hand is severely flexed at the wrist and the fingers are severely flexed at the interphalangeal joints.*

DUPUYTREN'S CONTRACTURE

Dupuytren's contracture *is caused by fibrosis and shortening of the palmar aponeurosis. Thickening and shortening of the bands of the aponeurosis over the flexor tendons results in flexion of the ring and little fingers.*

D. The **radial and ulnar arteries** arise from the brachial artery in the cubital fossa and supply the forearm and hand.

1. The **ulnar artery** supplies the medial side of the forearm and the hand.

a. The **common interosseous artery:**

(1) Arises from the ulnar artery in the cubital fossa.

(2) Branches into an anterior interosseous artery and a posterior interosseous artery, which supply the deep muscles in the anterior and posterior forearm, respectively.

b. The **superficial arch:**

(1) Is a continuation of the ulnar artery.

(2) Forms the **superficial palmar arch** in the palm of the hand and branches into palmar metacarpal arteries, which supply the hand and digits.

c. The **deep branch of the ulnar artery** anastomoses with the medial part of the **deep palmar arch.**

2. The **radial artery** supplies the lateral side of the forearm and the hand.

a. The **superficial branch** of the radial artery:

(1) Arises from the radial artery at the wrist.

(2) Anastomoses with the lateral part of the superficial palmar arch.

b. The **distal part of the radial artery:**

(1) Courses dorsal to the wrist and crosses the floor of the anatomic snuffbox.

(2) Enters the deep part of the palm after passing between the 2 heads of the first dorsal interosseous muscle.

(3) Forms the **deep palmar arch** that branches into dorsal metacarpal arteries, which supply the hand and digits.

III. Venous Drainage

A. The **cephalic** and **basilic** veins are superficial veins and arise from a venous network on the dorsal aspect of the hand.

1. The **cephalic vein** arises from the lateral aspect of the venous network, courses on the anterolateral aspect of the forearm and arm, passes between the deltoid and pectoralis major muscles, and drains into the axillary vein.

2. The **basilic vein** arises from the medial aspect of the venous network, courses on the medial aspect of the forearm, courses with the brachial artery in the arm, and joins with the brachial vein to form the axillary vein.

B. The deep veins of the upper limb course with arteries of the same name.

IV. Innervation (Figures 6–3 to 6–6, Table 6–8)

A. The upper limb is innervated by the **brachial plexus**, which is formed by an intermingling of ventral rami from the C5 through T1 spinal nerves.

B. The typical branching pattern of the ventral rami of the brachial plexus is **5-3-6-3-10-5**.

1. **Five ventral rami** from the C5 through T1 spinal nerves (also known as the 5 roots) intermingle to form **3 trunks.**

a. The ventral rami of C5 and C6 join to form the **superior trunk.**

b. The ventral ramus of C7 continues on as the **middle trunk.**

c. The ventral rami of C8 and T1 join to form the **inferior trunk.**

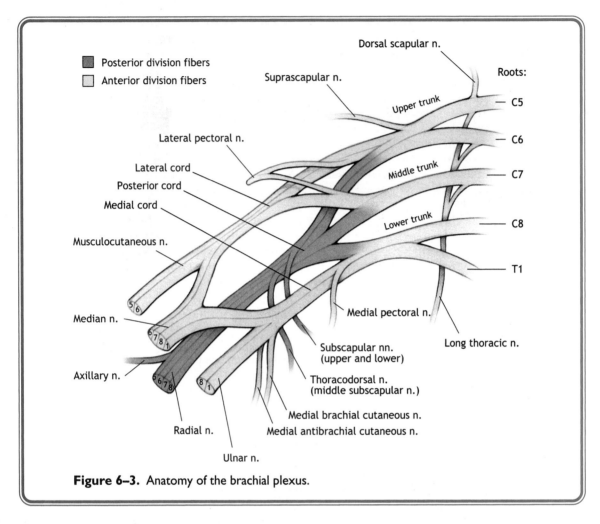

Figure 6–3. Anatomy of the brachial plexus.

GRADIENT OF INNERVATION

*The **ventral rami** of the brachial plexus exhibit a **proximal to distal gradient of innervation**.*
- *Nerves that contain fibers from the **superior rami** of the plexus (C5 and C6) innervate **proximal muscles** in the upper limb (eg, muscles that act at the pectoral girdle and shoulder).*
- *Nerves that contain fibers from the **inferior rami** of the plexus (C8 and T1) innervate **distal muscles** (eg, hand muscles).*
- *Nerves containing fibers from the **intermediate rami** of the plexus (C6 to C8) innervate muscles that act **mainly at the elbow and at the wrist**.*

     2. Six divisions are formed when each trunk divides into anterior division fibers and posterior division fibers.

SIGNIFICANCE OF ANTERIOR AND POSTERIOR DIVISION FIBERS

***Muscles in the anterior arm, anterior forearm, and hand** that act mainly as **flexors** are innervated by nerves that contain anterior division fibers. The musculocutaneous, ulnar, median, lateral, and medial pectoral nerves contain anterior division fibers.*

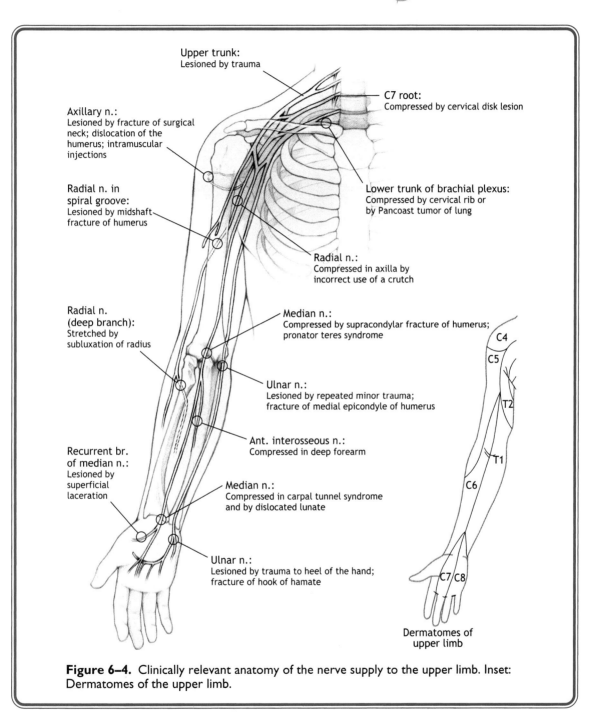

Upper trunk:
Lesioned by trauma

C7 root:
Compressed by cervical disk lesion

Axillary n.:
Lesioned by fracture of surgical neck; dislocation of the humerus; intramuscular injections

Radial n. in spiral groove:
Lesioned by midshaft fracture of humerus

Lower trunk of brachial plexus:
Compressed by cervical rib or by Pancoast tumor of lung

Radial n.:
Compressed in axilla by incorrect use of a crutch

Radial n. (deep branch):
Stretched by subluxation of radius

Median n.:
Compressed by supracondylar fracture of humerus; pronator teres syndrome

Ulnar n.:
Lesioned by repeated minor trauma; fracture of medial epicondyle of humerus

Ant. interosseous n.:
Compressed in deep forearm

Recurrent br. of median n.:
Lesioned by superficial laceration

Median n.:
Compressed in carpal tunnel syndrome and by dislocated lunate

Ulnar n.:
Lesioned by trauma to heel of the hand; fracture of hook of hamate

C4
C5
T2
T1
C6
C7/C8

Dermatomes of upper limb

Figure 6–4. Clinically relevant anatomy of the nerve supply to the upper limb. Inset: Dermatomes of the upper limb.

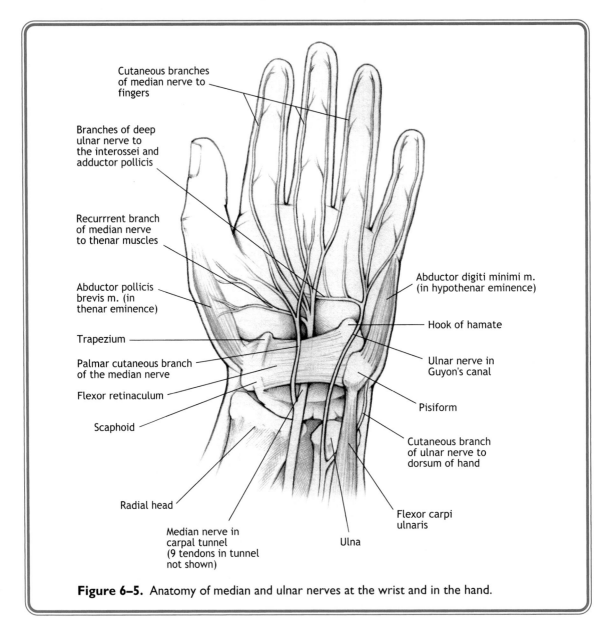

Figure 6–5. Anatomy of median and ulnar nerves at the wrist and in the hand.

Muscles in the posterior arm and posterior forearm that act mainly as extensors are innervated by nerves that contain posterior division fibers. The axillary, radial, upper, middle, and lower subscapular nerves contain posterior division fibers.

    3. Three cords are formed when the anterior fibers join together and when the posterior division fibers join together.

        a. The anterior division fibers of the superior and middle trunks join to form the **lateral cord**, and the anterior division fibers of the inferior trunk continue as the **medial cord**.

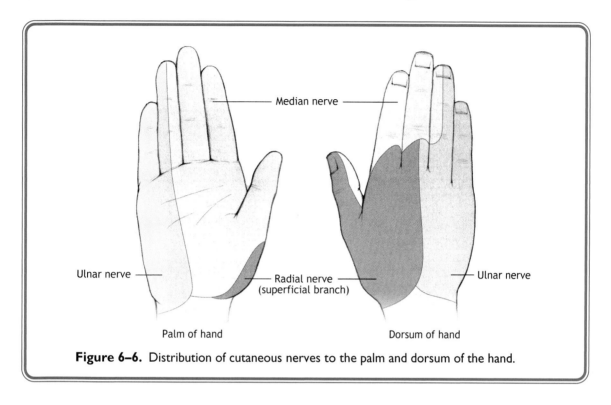

Figure 6–6. Distribution of cutaneous nerves to the palm and dorsum of the hand.

    b. The posterior division fibers of all 3 trunks unite to form the **posterior cord**. The cords are named for their relationship (lateral, medial, or posterior) to the axillary artery.

  4. **Ten collateral nerves** arise from the ventral rami, trunks, and cords before formation of the 5 terminal nerves. Each collateral nerve typically contains fibers from at least 2 adjacent ventral rami of the brachial plexus.

  5. **Five terminal nerves** arise from the 3 cords. Each terminal nerve typically contains fibers from at least 2 adjacent ventral rami of the brachial plexus.

    a. Three terminal nerves—the **musculocutaneous, ulnar, and median**—arise from the lateral and medial cords in the form of an M.

    b. Two terminal nerves—the **radial** and **axillary**—arise from the posterior cord.

PREPLEXUS INJURIES

Preplexus injuries affect the **ventral rami** or the **trunks** of the brachial plexus proximal to the formation and branching of terminal and collateral nerves and have a more widespread effect than lesions to individual collateral or terminal nerves.

ERB-DUCHENNE SYNDROME

The **Erb-Duchenne syndrome** results from a lesion of the **C5 and C6 ventral rami** in the **superior trunk** of the plexus (Figures 6–3 and 6–4, Table 6–8).

• **Proximal musculature** in the upper limb is mainly affected; thus, muscles acting at the shoulder and at the elbow will be weakened.

Table 6–8. Effects of lesions to major roots and nerves of brachial plexus.

Lesioned Root	Cause of Lesion	Location of Dermatome	Muscles Affected	Suppressed Reflex
C5	Upper trunk compression (Erb-Duchenne syndrome) Herniation of disk between C4 and c5	Posterior shoulder, upper lateral arm	Deltoid Rotator cuff Biceps Brachioradialis	Biceps tendon
C6	Upper trunk compression (Erb-Duchenne syndrome) Herniation of disk between C5 and c6	Lateral forearm, thenar eminence, dorsal and palmar sides of thumb and index finger	Biceps Brachioradialis Brachialis Supinator	Biceps tendon
C7	Cervical spondylosis Herniation of disk between C6 and C7	Midpalm, dorsal and palmar sides of middle finger	Latissimus dorsi Pectoralis major Triceps Wrist extensors	Triceps tendon
C8	Lower trunk compression by cervical rib or pancoast tumor (Klumpke's paralysis)	Hypothenar eminence, dorsal and palmar sides of ring and little finger	Finger flexors Wrist flexors Intrinsic hand muscles	
T1	Lower trunk compression by cervical rib or pancoast tumor (Klumpke's paralysis)	Medial arm and medial forearm to wrist	Intrinsic hand muscles	

Lesioned Nerve	Cause of Lesion	Altered Cutaneous Sensation	Weakness in	Common Sign of Lesion
Axillary (C5, C6)	Fracture of surgical neck of humerus, dislocated humeral head	Over deltoid insertion	Abduction at shoulder	
Radial (C5, C6, C7,C8)	Fracture midshaft of humerus Subluxation of radial head Dislocated humeral head Compression in axilla	Over first dorsal interosseous, anatomic snuffbox	Extension at wrist Extension of all fingers at Metacarpophalangeal joints Supination Thumb extension and abduction	Wrist drop
Median (C6, C7, C8, T1)	**Distal lesion:** Carpal tunnel syndrome Dislocated lunate	Lateral 3 1/2 digits	Opposition of thumb	Ape hand
	Proximal lesion: Fracture of supracondylar humerus Compression between heads of pronator teres	Lateral 3 1/2 digits, thenar eminence	Distal muscles and: Flexion at wrist Flexion of lateral fingers Flexion of thumb Pronation	Ulnar deviation of wrist Hand of benediction
Ulnar (C8, T1)	**Distal lesion:** Fracture of hook of hamate		Abduction and adduction of fingers, adduction of thumb Extension of fingers	Claw hand
	Proximal lesion: Fracture of or trauma at medial epicondyle of humerus Fracture of clavicle	Medial 1 1/2 digits, hypothenar eminence	Distal muscles and: Flexion of medial fingers Flexion at wrist	Radial deviation of wrist
Musculocuta-neous (C5, C6, C7)	Trauma Upper trunk compression	Lateral forearm	Flexion at elbow, supination	

There may be altered sensation in skin of the C5 and C6 dermatomes in the lateral arm, forearm, thumb and index finger (Figure 6–4 inset).
- In patients with this syndrome, the upper limb is held in a **"waiter's tip" position**, which results from a loss of abduction and a weakness of flexion and lateral rotation at the glenohumeral joint. The arm is adducted, extended, and medially rotated.
- Elbow flexion, supination, and wrist extension may also be weakened. The forearm is pronated, and the carpal flexors flex the hand at the wrist.
- The rhomboid muscles, the levator scapulae muscle, and the serratus anterior muscle are unaffected; the nerves that supply these muscles arise from ventral rami proximal to the lesion.

KLUMPKE'S PARALYSIS

Klumpke's paralysis results from compression of the **C8 and T1 ventral rami** in the **inferior trunk** of the plexus (Figures 6–3 and 6–4, Table 6–8).
- **Distal muscles** in the upper limb that are innervated by the C8 and T1 fibers in the lower trunk will be primarily affected.
There may be altered sensation in skin of the C8 and T1 dermatomes in the medial hand, ring and little fingers, and medial forearm (Figure 6–4 inset).
- Patients with this syndrome experience weakness of the intrinsic muscles of the hand innervated by C8 or T1 fibers in the median and ulnar nerves, resulting in a combination of an **ape hand** and a **claw hand**.
- Patients may have **difficulty making a fist** because of weakness in muscles that act to flex at the MP and interphalangeal joints of the fingers.
- Muscles that act at the shoulder and elbow are unaffected.
- Klumpke's paralysis may be caused by compression of the inferior trunk by a **cervical rib** or by a **Pancoast tumor** in the apical part of a lung. It may be seen in conjunction with other symptoms of thoracic outlet syndrome.

    C. **Ten collateral nerves** arise from the roots, trunks, and cords of the plexus.
        1. The **suprascapular nerve** (C5, C6):
            a. Arises from the upper trunk of the plexus and crosses the posterior triangle of the neck deep to the trapezius.
            b. Passes through the scapular notch under the superior transverse scapular ligament, through the supraspinatus fossa, around the spine of the scapula, and into the infraspinous fossa.
            c. Supplies the **supraspinatus** and the **infraspinatus** muscles.

SUPRASCAPULAR NERVE LESIONS

The **suprascapular nerve** may be compressed as it courses through the **scapular notch**. Patients with a lesion of the suprascapular nerve may experience shoulder pain, weakness in abduction of the arm at the glenohumeral joint and in lateral rotation at the glenohumeral joint.

        2. The **lateral pectoral nerve** (C5, C6, C7):
            a. Arises from the lateral cord.
            b. Supplies the **pectoralis major**, in particular, its clavicular head.
        3. The **medial pectoral nerve** (C8, T1):
            a. Arises from the medial cord.
            b. Supplies the **pectoralis minor,** then passes through it to innervate the pectoralis major muscle; the lateral and medial pectoral nerves are named because of their origin from the medial and lateral cords of the plexus, respectively.

4. The medial **brachial cutaneous nerve** (C8, T1):
  a. Arises from the medial cord.
  b. Supplies skin on the medial aspect of the arm.
5. The **medial antebrachial cutaneous** nerve (C8, T1):
  a. Arises from the medial cord just distal to the medial brachial cutaneous nerve.
  b. Supplies skin on the medial aspect of the forearm.
6. The **upper subscapular nerve** (C5, C6):
  a. Arises from the posterior cord.
  b. Supplies the upper part of the **subscapularis** muscle.
7. The **thoracodorsal nerve** (C6, C7, C8) (or middle subscapular nerve):
  a. Arises from the posterior cord between the upper and lower subscapular nerves.
  b. Supplies the **latissimus dorsi.**

THORACODORSAL NERVE LESIONS

*Surgical procedures of the axilla may result in a lesion of the **thoracodorsal nerve**. Patients may have difficulty in elevating the trunk (as if attempting to climb or do a pull-up) and may have difficulty in using a crutch.*

8. The **lower subscapular nerve** (C5, C6):
  a. Arises from the posterior cord.
  b. Supplies the lower part of **subscapularis muscle** and the **teres major muscle.**
9. The **dorsal scapular nerve** (C5):
  a. Arises from the C5 ventral ramus of the plexus.
  b. Courses through the substance of the scalenus medius muscle and crosses the posterior triangle of the neck.
  c. Passes deep to the vertebral border of the scapula.
  d. Supplies the **levator scapulae, rhomboid major, and rhomboid minor muscles.**
10. The **long thoracic nerve** (C5, C6, C7):
  a. Arises from the superior 3 ventral rami of the brachial plexus.
  b. Passes posterior to the C8 and T1 ventral rami and then courses superficially on the lateral thoracic wall with the lateral thoracic artery.
  c. Supplies the **serratus anterior** muscle.

LONG THORACIC NERVE LESIONS

*The **long thoracic nerve** is most commonly injured as it courses superficial to the serratus anterior on the lateral wall of the thorax.*

• *Patients with a lesion of the long thoracic nerve cannot hold the vertebral border of the scapula flat against the back and may have a **"winging"** of the vertebral border of the scapula (**medial winged scapula**).*
• *Patients also experience weakness in the ability to protract the scapula and difficulty in raising their arm above their head.*

V. The 5 Terminal Nerves of the Plexus

A. Musculocutaneous Nerve (C5, C6, and C7)
  1. The musculocutaneous nerve contains anterior division fibers from the superior 3 ventral rami of the brachial plexus.
  2. It continues from the lateral cord and passes through the coracobrachialis muscles into the anterior compartment of the arm.

3. The musculocutaneous nerve courses between the biceps and the brachialis muscles, emerging just lateral to the tendon of the biceps.

4. It supplies the coracobrachialis, biceps brachii, and brachialis muscles.

5. It continues as the lateral antebrachial cutaneous nerve, which supplies skin of the lateral aspect of the forearm.

MUSCULOCUTANEOUS NERVE LESIONS (TABLE 6–8)

Lesions of the musculocutaneous nerve are uncommon.

• *The nerve may be compressed as it passes through the coracobrachialis muscle.*

• *Patients with such a lesion may experience weakness in flexion of the forearm at the elbow and weakness in supination.*

    B. Median Nerve (C6, C7, C8, T1)

       1. The median nerve contains anterior division fibers from all of the ventral rami of the brachial plexus except for C5.

       2. Fibers from the **lateral and medial cords** unite anterior to the axillary artery to form the median nerve.

       3. The median nerve courses medial to the tendon of the biceps and the brachial artery in the arm. It reaches the forearm after passing anterior to the elbow and deep to the bicipital aponeurosis.

       4. In the forearm, the nerve passes between the **2 heads of the pronator teres** and then courses between the flexor digitorum superficialis and flexor digitorum profundus muscles to reach the wrist.

       5. The median nerve supplies **superficial muscles in the anterior forearm**, including the flexor carpi radialis, palmaris longus, flexor digitorum superficialis, and pronator teres muscles.

       6. It gives rise to the anterior interosseous nerve in the cubital fossa, which supplies deep muscles in the anterior forearm, including the lateral half of the flexor digitorum profundus, the flexor pollicis longus, and the pronator quadratus. The **only muscles in the anterior forearm not supplied by the median nerve** are the brachioradialis (radial nerve) flexor carpi ulnaris and the medial half of the flexor digitorum profundus (ulnar nerve).

       7. The median nerve passes through the carpal tunnel at the wrist between the flexor retinaculum and the flexor tendons (Figure 6–5).

       8. It gives off a **recurrent branch** distal to the carpal tunnel, which supplies the 3 thenar muscles.

       9. The median nerve gives off **muscular branches** distal to the carpal tunnel, which supply the lumbricals to the index and middle fingers.

     10. It also gives off **common palmar digital** nerves and **proper digital cutaneous nerves**, which supply the palmar aspects and sides of the radial 3 1⁄2 digits (thumb, index, and middle fingers and half of the ring finger), and the nail beds of 2 1⁄2 digits (Figures 6–5 and 6–6).

     11. Finally, the median nerve gives off a **palmar cutaneous branch** proximal to the carpal tunnel, which crosses superficial to the flexor retinaculum and supplies skin of the palm up to the base of the lateral digits.

MEDIAN NERVE LESIONS (TABLE 6–8)

*In carpal tunnel syndrome, the **median nerve** is compressed as it courses through the carpal tunnel between the **flexor tendons** and the **flexor retinaculum** (see Figure 6–4).*

- Patients with **carpal tunnel syndrome** experience numbness and pain, particularly at night, over the palmar aspects of the thumb, index, and middle fingers.
- **Cutaneous sensation** from the **lateral aspect** of the palm may be spared because the palmar branch of the median nerve does not traverse the carpal tunnel.
- **Weakness of the thenar muscles** may be evident and results in an ape hand, where the thumb cannot be opposed, and is adducted and extended.
- The **lateral 2 lumbricals** may be weakened, resulting in a slight clawing of the index and middle fingers because of reduced ability to flex the MP joints and extend the interphalangeal joints of these digits.
- The **median nerve** may be compressed proximal to the cubital fossa in a supracondylar fracture of the humerus, ordistal to the cubital fossa, as it passes between the 2 heads of the pronator teres (Figure 6–4).
- In addition to altered sensation in the lateral part of the hand and a loss of thumb opposition, these patients experience weakness in pronation and weakness in the ability to flex the thumb, the PIP and DIP joints of the index and middle fingers, and the PIP joints of the ring and little fingers.
- Patients with either of these lesions may have a **"hand of benediction,"** in which the index and middle fingers remain extended when the patient attempts to flex those digits to make a fist.
- The **recurrent branch of the median nerve** may be lesioned distal to the carpal tunnel as a result of laceration of the nerve adjacent to the thenar eminence (see Figure 6–5).
- A lesion of the **recurrent branch** affects the thenar muscles, resulting in an ape hand, with no cutaneous deficits.

ANTERIOR INTEROSSEOUS NERVE LESIONS

CLINICAL
CORRELATION

The **anterior interosseous nerve** may be compressed near the interosseous membrane deep in the anterior forearm (Figure 6–4).
- A lesion of the anterior interosseous nerve may result in a weakness of pronation (pronator quadratus) and weakness in flexion at the index and middle fingers at the DIP joints.
- Patients with a lesion of the anterior interosseous nerve may have a weakness in the ability to flex the distal phalanx of the thumb (flexor pollicis longus) and an inability to form the letter o by touching the tip of the thumb to the tip of the index finger.

    C. Ulnar Nerve (C8, T1)
      1. The ulnar nerve consists of anterior division fibers of the C8 and T1 ventral rami, which continue from the medial cord.
      2. It courses medially in the arm and reaches the forearm after passing posterior to the elbow through a groove between the olecranon and the medial epicondyle of the humerus.
      3. The ulnar nerve passes between the 2 heads of the flexor carpi ulnaris in the anterior forearm, and then courses between the flexor carpi ulnaris and the flexor digitorum profundus muscles to reach the wrist.
      4. It innervates 1 1/2 muscles in the anterior forearm: the flexor carpi ulnaris muscle and the medial half of the flexor digitorum profundus muscle.
      5. The ulnar nerve gives rise to a palmar cutaneous branch, which passes anterior to the flexor retinaculum to supply the medial aspect of the palm of the hand.
      6. It also gives rise to a dorsal cutaneous branch. This branch divides into dorsal digital nerves, which supply the skin of the dorsum of the hand at the bases of the little finger and the medial side of the ring finger.
      7. The ulnar nerve passes through the canal of Guyon between the pisiform bone and the hook of the hamate and anterior to the carpal tunnel (see Figure 6–5).

8. It divides into a superficial branch that innervates the palmaris brevis and the skin of the medial side of the palm and dorsum of the hand, the dorsal and palmar aspects of the little finger, and the medial half of the ring finger (see Figures 6–5 and 6–6).

9. The ulnar nerve divides into a deep branch that innervates the 3 muscles of the hypothenar eminence and the 2 lumbricals to the ring and little finger (Figures 6–4 to 6–6). This branch passes deep into the palm between the flexor digiti minimi and the abductor digiti minimi muscles and innervates the adductor pollicis and the 7 interosseous muscles.

ULNAR NERVE LESIONS (TABLE 6–8)

The **ulnar nerve** may be compressed at the wrist as it passes between the hook of the hamate and the pisiform bone in Guyon's canal, or it may be lesioned as a result of a fracture of the hook of the hamate (Figure 6–5).

- Patients with an ulnar nerve lesion at the wrist may have an ulnar claw hand, which is caused by a weakness of the medial 2 lumbricals that flex at the MP joints and extend at the interphalangeal joints of the ring and little fingers.

- Patients also experience weakness in the ability to abduct or adduct fingers or adduct the thumb at the MP joints (interosseous muscles and adductor pollicis). They are unable to hold a piece of paper between the thumb and index finger or between adjacent fingers.

- **Weakness of the interosseus muscles** may also result in a slight clawing of the index and middle fingers (the lateral 2 lumbricals, which are innervated by the median nerve, are unaffected).

- The muscles in the **hypothenar eminence** may also be affected; patients experience weakness in flexion, abduction, and opposition of the fifth finger.

- Altered sensation in skin of the medial aspect of the hand and medial digits may be evident.

- The **ulnar nerve is most commonly lesioned** at the elbow as it courses adjacent to the medial epicondyle of the humerus, or it may be compressed between the 2 heads of the flexor carpi ulnaris (cubital tunnel syndrome) muscle (see Figure 6–4).

- In addition to an ulnar claw hand and weakness in abduction and adduction of the digits, patients may experience a weakness in the ability to flex the DIP joints of the ring and little fingers and a weakness in the ability to flex the hand at the wrist.

- Patients with an ulnar nerve lesion at these sites have pain and paresthesia in the medial 1 1/2 digits.

    D. Axillary Nerve (C5, C6)

      1. The axillary nerve consists of posterior division fibers of the C5 and C6 ventral rami in the posterior cord.

      2. It courses through the quadrangular space around the surgical neck of the humerus.

      3. The axillary nerve innervates the deltoid and the teres minor muscles.

      4. It innervates the skin of the arm covering the attachment of the deltoid to the humerus.

AXILLARY NERVE LESIONS (TABLE 6–8)

The **axillary nerve** may be injured as a result of a dislocation of the head of the humerus from the glenoid fossa or by a fracture of the surgical neck of the humerus (see Figure 6–4).

- Patients may experience weakness in the ability to abduct the arm at the glenohumeral joint because of loss of the deltoid. In addition, there may be altered sensation in skin covering the deltoid.

- The deltoid may undergo atrophy, resulting in a loss of the rounded contour of the shoulder.
- There may be weakness in lateral rotation because of weakness of the teres minor muscle.

E. Radial Nerve (C5, C6, C7, C8, T1)

1. The **radial nerve** consists of posterior division fibers from the posterior cord.
2. It crosses the tendon of the latissimus dorsi posterior to the axillary artery and then courses around the posterior aspect of the shaft of the humerus between the medial and lateral heads of the triceps.
3. The radial nerve innervates the 3 heads of the triceps brachii muscle, the brachioradialis and the extensor carpi radialis longus muscles, and the skin of the posterior arm.
4. It enters the forearm anterior to the lateral epicondyle of the humerus.
5. In the proximal part of the forearm, the radial nerve divides into a superficial cutaneous branch and a deep muscular branch.
   a. The superficial branch of the radial nerve innervates skin over the lateral side of forearm, the lateral side of the dorsal aspect of the hand, and the dorsal aspect of the lateral 3½ digits to the PIP joints.
   b. The deep branch passes posteriorly around the proximal part of the radius within the supinator muscle and into the posterior forearm. It innervates muscles of the posterior forearm, including the supinator, extensor digitorum, extensor carpi radialis brevis, extensor carpi ulnaris, extensor pollicis longus, extensor pollicis brevis, abductor pollicis longus, extensor digiti minimi, and extensor indicis muscles.

RADIAL NERVE LESIONS

The radial nerve is commonly lesioned as a result of a **spiral fracture** of the midshaft of the humerus (see Figure 6–4).

- A common feature of this type of lesion is **"wristdrop,"** or weakness in the ability to extend the hand at the wrist, and a loss of extension at the MP joints of all digits.
- Supination may be weakened but not lost; the biceps brachii (the other supinator), which is innervated by the musculocutaneous nerve, will be unaffected.
- Extension of the forearm at the elbow is spared because the triceps receives its innervation proximal to the fracture.
- Patients with radial nerve lesions may experience pain and paresthesia in skin over the first dorsal interosseous muscle between the thumb and the index finger.
- Distal to the elbow, the **deep branch of the radial nerve** may incur a lesion as it courses through the supinator by a **subluxation** of the head of the radius. Patients with this type of lesion may experience wristdrop and weakness in the ability to extend the MP joints but no sensory deficits.

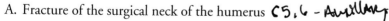

CLINICAL PROBLEMS

Match the deficit in Questions 1–10 with the appropriate injury in Choices A–G. (Choices may be used once, more than once, or not at all.)

 A. Fracture of the surgical neck of the humerus C5,6 - Auxillary

 B. Fracture of the hook of the hamate - Ulnar

C. Fracture of the scaphoid

D. A midshaft fracture of the humerus *- Radial*

E. A subluxation of the head of the radius *Axillary Radius*

F. Anterior dislocation of the lunate *Median*

G. Supracondylar fracture of the humerus *- Median*

1. Patient has a loss of sensation in skin over the anatomic snuffbox, and "wristdrop." *Radial*

2. Patient has atrophy of the muscles of the thenar eminence and paresthesia in the lateral digits; wrist and finger flexion are intact. *Median*

3. Patient has weakness in the ability to extend the distal phalanges at the interphalangeal joints of the ring and little fingers, and wasting of the hypothenar eminence. *Ulnar*

4. Patient has a loss of pronation, ulnar deviation of the wrist during wrist flexion, and an "ape hand." *Median*

5. Patient has weakness in the ability to abduct the arm at the shoulder, and some weakness in the ability to laterally rotate the arm at the shoulder. *Axillary*

6. Patient has altered sensation in skin of the palmar surface of the lateral three-and-a-half digits; flexion of the digits is intact. *Median*

7. Patient has weakness in the ability to extend the hand at the wrist; cutaneous sensation is intact. *Radial*

8. Patient cannot flex either the distal phalanx of the thumb at the interphalangeal joint or the fingers at the proximal interphalangeal joints, and has altered sensation in skin overlying the thenar eminence. *Median*

9. Patient has weakness in the ability to extend the thumb, and has altered sensation in skin overlying the first dorsal interosseous muscle. *Radial*

10. Patient has suffered an injury that has also lacerated the profunda brachial artery. *Radial*

A 19-year-old woman was thrown while riding a bicycle. She attempted to break her fall with an outstretched hand and suffered a fracture. In the emergency room, an examination reveals an inability to extend the hand at the wrist.

11. What might have been the site of a fracture that caused the muscle weakness?

   A. Clavicle

   B. Hook of the hamate

   C. Styloid process of the radius

   D. Midshaft of the humerus

   E. Scaphoid

In the patient in Question 11, a hematoma develops in the area of the fracture.

12. What blood vessel might have also been lacerated at the fracture site?

   A. Subscapular artery

   B. Posterior circumflex humeral artery

C. Profunda brachial artery

D. Radial artery

E. Ulnar artery

A 35-year-old woman comes to the clinic complaining of pain radiating down the medial aspect of the left forearm and into the medial aspect of the left hand. She states that her left hand is weaker than her right hand. You note that her thenar and hypothenar eminences are smaller in the left hand compared with the right, and her radial pulse is diminished on the left.

13. Compression of what neural structure might account for the patient's symptoms?

A. Upper trunk of the brachial plexus

B. Median nerve

C. Ulnar nerve

D. Lower trunk of the brachial plexus

E. Posterior cord of the brachial plexus

Your patient suffers from compression of a nerve. The patient has weakness of pronation and flexion at the index and middle fingers at the distal interphalangeal joints and an inability to form the letter *o* by touching the tip of the thumb to the tip of the index finger. There are no sensory deficits.

14. What nerve may have been compressed?

A. Recurrent branch of the median nerve

B. Deep branch of the radial nerve

C. Deep branch of the ulnar nerve

D. Anterior interosseous branch of the median nerve

E. Superficial branch of the radial nerve

A 36-year-old woman suffers a traumatic injury to the upper limb that lesions a nerve. The lesion results in an inability to spread and extend her fingers, and a "clawing" of the ring and little fingers.

15. What two spinal cord segments contribute to the nerve that is damaged?

A. C8 And T1

B. C7 and C8

C. C6 and C7

D. C5 and C6

E. C4 and C5

Your patient has been thrown from a motorcycle and suffers trauma to the upper limb. In the hospital, the left arm of the patient hangs at his side because of a loss of abduction and a weakness of flexion and lateral rotation at the glenohumeral joint.

16. What else might you expect to observe in the patient?

A. Atrophy of the hypothenar eminence

B. Weakness in the ability to protract the scapula

C. Weakness in supination

D. Inability to abduct and adduct the fingers

E. Altered sensation in skin in the medial aspect of the forearm

A patient suffers a fracture of the supracondylar part of the humerus, which compresses a nerve and an accompanying artery.

17. What might you observe in the patient?

　A. Clawing of the ring and little fingers

　B. Altered sensation in skin over the anatomic snuffbox

　C. Inability to extend the thumb

　D. Dupuytren's contracture

　E. A hand of benediction

Your patient suffers from a progressive compression of the axillary artery posterior to the pectoralis minor.

18. Collateral circulation develops, bypassing the blockage by way of anastomosis between the suprascapular artery and what other artery?

　A. Dorsal scapular artery

　B. Profunda brachial artery

　C. Thoracoacromial artery

　D. Subscapular artery

　E. Radial artery

Your patient has radial deviation of the hand at the wrist when he attempts to flex the wrist and altered sensation in the skin covering the hypothenar eminence.

19. What might account for the symptoms?

　A. Fracture of surgical neck of the humerus

　B. Fracture of the distal end of the radius

　C. Anterior and inferior dislocation of the head of the humerus

　D. Fracture of the scaphoid bone

　E. Fracture of the medial epicondyle of the humerus

The same patient as in Question 19 develops a significant "clawing" of the fifth digit secondary to the nerve injury.

20. What muscle has been weakened that causes the clawing?

　A. Dorsal interosseous

　B. Extensor digitorum

　C. Lumbrical

　D. Flexor digitorum superficialis

　E. Flexor digiti minimi

A patient has suffered a fracture of the surgical neck of the humerus.

21. What muscle might be weakened?

A. Deltoid

B. Supraspinatus

C. Biceps brachii

D. Teres major

E. Latissimus dorsi

A man who works as a cartoonist for a living begins to develop pain and paresthesia in his right hand at night. The altered sensation is most evident on the palmar aspects of the index and middle fingers.

22. What else might you expect to see in the patient?

A. Atrophy of the thenar eminence

B. Weakness in the ability to extend the thumb

C. Radial deviation of the hand at the wrist during wrist flexion

D. Altered sensation in skin over the anatomic snuffbox

E. Inability to spread and oppose the fingers

The middle trunk of the brachial plexus is lesioned.

23. Axons in all of the following nerves will be affected except the

A. Median nerve

B. Ulnar nerve

C. Musculocutaneous nerve

D. Radial nerve

E. Thoracodorsal nerve

A 28-year-old secretary develops pain and numbness in her thumb and index and middle fingers at night.

24. Which of the following motor functions may also be weakened?

A. Adduction of the thumb

B. Extension of the thumb

C. Flexion of the index finger at the interphalangeal joints

D. Abduction of the thumb

E. Pronation of the forearm

ANSWERS

1. D

2. F

3. B

4. G

5. A

6. F

7. E

8. G

9. D

10. D

11. The answer is D. The radial nerve is commonly lesioned as a result of a spiral fracture of the midshaft of the humerus.

12. The answer is C. The deep brachial artery courses with the radial nerve around the humerus.

13. The answer is D. Klumpke's paralysis results from compression of the C8 and T1 ventral rami in the inferior trunk of the plexus at the level of the first rib.

14. The answer is D. A lesion of the anterior interosseous branch of the median nerve would account for the symptoms.

15. The answer is A. The patient has a lesion of the ulnar nerve.

16. The answer is C. Weakness in supination may result from compression of the C5 and C6 fibers in the upper trunk. These fibers innervate the biceps brachii by way of the musculocutaneous nerve.

17. The answer is E. Patients with a lesion of the median nerve in a supracondylar fracture may have a "hand of benediction," in which the index and middle fingers remain extended when the patient attempts to flex those digits while making a fist.

18. The answer is D. Anastomoses develop superior and posterior to the scapula between the thoracodorsal and circumflex scapular branches of the subscapular artery and the suprascapular, dorsal scapular, and posterior intercostal arteries.

19. The answer is E. Fracture of the medial epicondyle of the humerus may result in a lesion to the ulnar nerve, causing the symptoms in this patient.

20. The answer is C. Weakness of the lumbricals innervated by the ulnar nerve results in the clawing of the digits. A dorsal interosseous muscle does not act on the fifth digit.

21. The answer is A. The deltoid is innervated by the axillary nerve.

22. The answer is A. The patient has carpal tunnel syndrome. Choices B and D result from radial nerve lesions. Choice E results from an ulnar nerve lesion. Choice C results from a median nerve lesion at the wrist.

23. The answer is B. The ulnar nerve contains only C8 and T1 fibers; all of the other answer choices contain C7 fibers form the middle trunk of the plexus.

24. The correct answer is B. The patient has carpal tunnel syndrome which may affect the thenar muscles that permit opposition by abducting, flexing and opposing the thumb.

CHAPTER 7
LOWER LIMB

I. Bones and Joints

A. The **pelvic girdle** consists of the ilium, ischium, and pubis, which are fused together at the acetabulum. The joints of the pelvic girdle include the sacroiliac joints and the hip joints (Figure 7–1).

    1. The pelvic girdle suspends the lower limb away from the axial skeleton to increase range of movement.

    2. Unlike the pectoral girdle, the pelvic girdle has **limited mobility**.

B. The **sacroiliac joints** are **gliding joints** formed between the wings of the ilia and the sacrum. The posterior sacroiliac ligaments resist rotation at the sacroiliac joints.

C. The **hip joint** is formed by the head of the femur and the acetabulum.

    1. The hip joint is a **ball-and-socket joint**, permitting flexion and extension, abduction and adduction, and medial and lateral rotation (Table 7–1).

    2. The **articular capsule** of the hip joint is strengthened by 3 ligaments that extend from bones of the pelvic girdle to the neck of the femur.

        a. The **iliofemoral ligament** is the strongest ligament at the hip joint, reinforces the anterior part of the capsule, and resists extension.

        b. The **pubofemoral ligament** reinforces the anterior and inferior parts of the capsule and resists abduction.

        c. The **ischiofemoral ligament** reinforces the posterior part of the capsule and resists extension by screwing the head of the femur into the acetabulum.

FRACTURE OF THE NECK OF THE FEMUR

*The **neck of the femur** is a common site of a fracture.*

- *In such a fracture, the head of the femur may undergo **avascular necrosis** as a result of disruption of branches of the medial circumflex femoral artery, the main source of arterial blood supply to the head and neck of the femur.*

In patients with fractures of the femoral neck, the thigh is laterally rotated by the short lateral rotators of the thigh at the hip and by the gluteus maximus.

DISLOCATION OF THE HEAD OF THE FEMUR

*A **dislocation of the head of the femur** at the hip joint occurs most commonly in the posterior direction.*

- *The thigh is shortened and medially rotated by the gluteus medius and minimus muscles.*

- *The sciatic nerve may be compressed, resulting in weakness of muscles in the posterior thigh, leg, and foot and paresthesia over the posterior and lateral parts of the leg and the dorsal and plantar surfaces of the foot.*

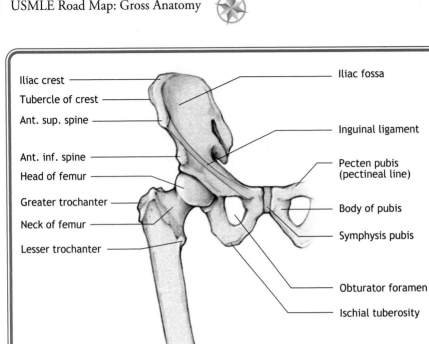

Iliac crest

Tubercle of crest

Ant. sup. spine

Ant. inf. spine

Head of femur

Greater trochanter

Neck of femur

Lesser trochanter

Femur

Patella

Lateral epicondyle

Lateral condyle

Head

Neck

Fibula

Iliac fossa

Inguinal ligament

Pecten pubis
(pectineal line)

Body of pubis

Symphysis pubis

Obturator foramen

Ischial tuberosity

Adductor tubercle

Medial epicondyle

Medial condyle

Tibial tuberosity

Tibia

Figure 7–1. Bones of the lower limb.

D. The knee joint is the largest synovial joint in the body and is formed by the lateral and medial condyles of the femur and the tibial plateaus (see Figure 7–1).

1. The knee joint is a **modified hinge joint**, which permits flexion and extension, rotation, and gliding (Table 7–2).

2. The knee is **strengthened by ligaments** of the articular capsule and by ligaments inside the articular capsule.

a. The **patellar ligament** is an extension of the quadriceps tendon, which strengthens the anterior and lateral parts of the capsule.

Table 7–1. Muscles that act at the hip joint.

Action	Muscles Involved	Innervation	Major Segments of Innervation
Flexion of femur	Iliacus and psoas major	Lumbar ventral rami	L2, L3
	Rectus femoris	Femoral	L2, L3, L4
	Sartorius	Femoral	L2, L3
	Tensor fasciae latae	Superior gluteal	L4, L5, S1
	Pectineus	Femoral	L2, L3
Extension of femur	Gluteus maximus	Inferior gluteal	L5, S1, S2
	Semimembranosus	Sciatic (tibial)	L5, S1
	Semitendinosus	Sciatic (tibial)	L5, S1, S2
	Biceps femoris, long head	Sciatic (tibial)	S1, S2
	Adductor magnus, ischial part	Obturator	L3, L4
Adduction of femur	Adductors longus, brevis, and magnus	Obturator	L2, L3, L4
	Gracilis	Oburator	L2, L3, L4
Abduction of femur	Gluteus medius and minimus	Superior gluteal	L4, L5, S1
	Tensor fasciae latae	Superior gluteal	L4, L5, S1
Medial rotation of femur	Gluteus minimus	Superior gluteal	L5, L5, S1
	Gluteus medius, anterior fibers	Superior gluteal	L4, L5, S1
Lateral rotation of femur	Gluteus maximus	Inferior gluteal	L5, S1, S2
	Sartorius	Femoral	L2, L3, L4
	Obturator internus and superior gemellus	Nerve to obturator internus	L5, S1, S2
	Obturator externus	Obturator	L3, L4
	Quadratus femoris and inferior gemellus	Nerve to quadratus femoris	L4, L5, S1
	Piriformis	Nerve to piriformis	L5, S1, S2

 b. The **oblique popliteal ligament** (part of the semimembranosus tendon) and the arcuate ligament strengthen the posterior part of the capsule.

 c. The **fibular and tibial collateral ligaments** support the lateral and medial parts of the joint, respectively.

  (1) The fibular collateral ligament extends from the lateral epicondyle of the femur to attach to the head of the fibula.

  (2) The tibial collateral ligament extends from the medial epicondyle of the femur to attach to the medial aspect of the tibia. The deep fibers of the tibial collateral ligament are attached to the medial meniscus.

  (3) The fibular and tibial collateral ligaments are most taut when the knee is fully extended.

Table 7–2. Muscles that act at the knee joint.

Action	Muscles Involved	Innervation	Major Segments of Innervation
Flexion of tibia and fibula	Semimembranosus	Sciatic (tibial)	L5, S1
	Semitendinosus	Sciatic (tibial)	L5, S1, S2
	Biceps femoris:		
	Long head	Sciatic (tibial)	S1, S2
	Short head	Sciatic (common peroneal)	L5, S1, S2
	Gracilis	Obturator	L2, L3, L4
	Sartorius	Femoral	L2, L3
	Popliteus	Tibial	L4, L5, S1
	Gastrocnemius	Tibial	S1, S2
Extension of tibia and fibula	Quadriceps femoris:		
	Vastus medialis	Femoral	L2, L3, L4
	Vastus lateralis	Femoral	L2, L3, L4
	Vastus intermedius	Femoral	L2, L3, L4
	Rectus femoris	Femoral	L2, L3, L4
Lateral rotation of tibia and fibula	Gluteus maximus	Inferior gluteal	L5, S1, S2
	Biceps femoris	Sciatic (tibial and common peroneal)	L5, S1, S2
	Tensor fasciae latae	Superior gluteal	L4, L5, S1
Medial rotation of tibia and fibula	Popliteus ("unlocks" extended knee)	Tibial	L4, L5, S1
	Semimembranosus	Sciatic (tibial)	L5, S1
	Semitendinosus	Sciatic (tibial)	L5, S1, S2
	Gracilis	Obturator	L2, L3, L4
	Sartorius	Femoral	L2, L3

     (4) The fibular and tibial collateral ligaments limit abduction and adduction of the knee when the knee is flexed.

  d. The **anterior cruciate ligament (ACL) and posterior cruciate ligament (PCL)** are intracapsular ligaments that help stabilize the joint.

  e. The ACL and PCL are named for the parts of the superior surface of the tibia to which they attach.

     (1) The ACL is the **"APEX"** ligament; it attaches to the *A*nterior aspect of the tibia and courses *P*osteriorly and *EX*ternally (laterally) to attach to the lateral condyle of the femur.

     (2) The ACL is weaker than the PCL, is most taut when the knee is extended, and resists hyperextension by preventing anterior displacement of the tibia on the femur (Figure 7–2).

     (3) The PCL is the **"PAIN"** ligament; it attaches to the *P*osterior aspect of the tibia and courses *A*nteriorly and *IN*ternally (medially) to attach to the medial condyle of the femur.

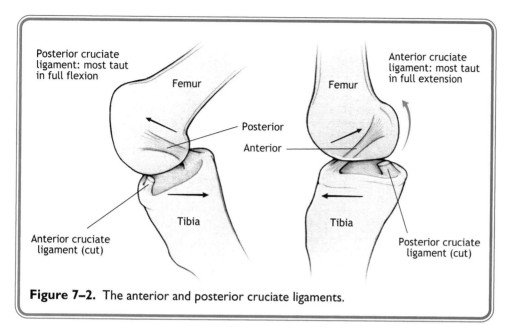

Figure 7–2. The anterior and posterior cruciate ligaments.

    (4) The PCL is most taut when the knee is flexed and resists excessive flexion by preventing posterior displacement of the tibia on the femur (see Figure 7–2).

  f. The **medial and lateral menisci** are intracapsular, crescent-shaped fibrocartilages that attach to the tibia and act as shock absorbers.

    (1) The medial meniscus has the shape of the letter *C.* It is less mobile than the lateral meniscus and is attached to the deep fibers of the tibial collateral ligament.

    (2) The lateral meniscus has the shape of the letter *o.* It is more mobile than the medial meniscus and is separated from the fibular collateral ligament by the tendon of the popliteus.

KNEE INJURIES

*The 3 **most commonly injured structures at the knee** are the tibial collateral ligament, the medial meniscus, and the ACL **(the terrible triad).***

- *A blow to the lateral aspect of the knee when the foot is on the ground may sprain the tibial collateral ligament; the attached medial meniscus may also be torn.*
- *Patients with a **medial meniscus tear** have pain when the leg is medially rotated at the knee.*
- ***ACL tears** may occur when the tibial collateral ligament and medial meniscus are injured; a blow to the anterior aspect of the flexed knee may tear only the ACL.*
- *Patients with a torn ACL exhibit an anterior drawer sign, in which the tibia may be displaced anteriorly from the femur in the flexed knee.*

    3. The fibula articulates with the tibia and the talus and does not articulate directly at the knee joint.

   E. The **ankle joints** include the **talocrural, subtalar, and transverse tarsal joints** (Figure 7–3).

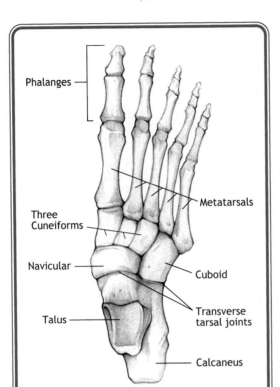

Figure 7–3. The bones of the ankle and foot.

1. Talocrural joint
   a. The **talocrural joint** is formed by an articulation between the trochlea of the talus and the lateral and medial malleoli of the fibula and tibia, respectively.
   b. This joint is a **hinge joint**, which permits dorsiflexion and plantar flexion of the foot (Table 7–3).
   c. The foot is more stable in a **dorsiflexed position** than in a plantar-flexed position because the anterior part of the superior surface of the trochlea of the talus is wider than the posterior part at the talocrural joint.
   d. The tendons of the tibialis posterior, flexor digitorum longus, and flexor hallucis longus muscles enter the sole of the foot after passing posterior and inferior to the medial malleolus.
   e. The tendons of the fibularis (peroneus) longus and fibularis (peroneus) brevis muscles enter the sole of the foot after passing posterior and inferior to the lateral malleolus.
   f. The **deltoid (medial) and the lateral ligaments** strengthen the talocrural joint.
      (1) The **deltoid ligament** is the stronger of the 2 collateral ligaments and has 4 components, which extend from the tibia to the talus, navicular, and calcaneus.

Table 7–3. Muscles that act at the talocrucal joint.

Action	Muscles Involved	Innervation	Major Segments of Innervation
Plantar flexion of foot	Gastrocnemius	Tibial	S1, S2
	Soleus	Tibial	S1, S2
	Plantaris	Tibial	L5, S1
	Tibialis posterior	Tibial	L5, S1
	Flexor digitorum longus	Tibial	L5, S1
	Flexor hallucis longus	Tibial	L5, S1, S2
Dorsiflexion of foot	Tibialis anterior	Deep peroneal	L4, L5
	Extensor hallucis longus	Deep peroneal	L5, S1
	Extensor digitorum longus	Deep peroneal	L4, L5, S1

(2) The **lateral ligament** has 3 components, which extend from the fibula to the talus and calcaneus.

ANKLE SPRAINS

Inversion ankle sprains are more common than eversion sprains at the talocrural joint. The anterior talofibular part of the lateral ligament is commonly torn in inversion ankle sprains.

  2. Subtalar joint
  a. The **ball-and-socket subtalar joint** is formed by the articulation between the talus and calcaneus.
  b. It permits supination and pronation.
    (1) **Supination** is a combination of plantar flexion, inversion, and adduction.
    (2) **Pronation** is a combination of dorsiflexion, eversion, and abduction.
  3. Transverse tarsal joints
  a. The transverse tarsal joints are **formed by the articulations of the talus with the navicular and the calcaneus with the cuboid.**
  b. The transverse tarsal joints **contribute to inversion and eversion** with the subtalar joint (Table 7–4).
 F. The **joints of the toes and the actions of muscles** of the foot are similar to those in the hand (Table 7–5).
  1. The **sole of the foot** contains short abductors and flexors of the great toe and little toe, respectively, but lacks eminences and opponens muscles.
  2. The **quadratus plantae** is a plantar muscle with no counterpart in the hand; it acts to straighten out the oblique pull of the flexor digitorum longus tendons.

II. Arterial Blood Supply (Figures 7–4a, 7–4b)

 A. The **femoral, popliteal, and anterior and posterior tibial arteries** provide most of the arterial blood supply to the lower limb.
  1. The **femoral artery**
   a. The femoral artery begins at the inguinal ligament as a continuation of the external iliac artery.

Table 7–4. Muscles that act at the transverse tarsal and subtalar joints.

Action	Muscles Involved	Innervation	Major Segments of Innervation
Inversion of foot	Tibialis anterior	Deep peroneal	L4, L5
	Extensor hallucis longus	Deep peroneal	L5, S1
	Tibialis posterior	Tibial	L5, S1
Eversion of foot	Peroneus longus	Superficial peroneal	L5, S1
	Peroneus brevis	Superficial peroneal	L5, S1
	Peroneus tertius	Deep peroneal	L4, L5, S1

 b. The femoral artery courses lateral to the femoral vein and medial to the femoral nerve through the femoral triangle in the anterior thigh.

 c. It enters the adductor canal and becomes the popliteal artery after passing through the adductor hiatus.

  (1) The profunda femoral artery gives rise to the lateral and medial circumflex arteries, which supply the thigh, the head and neck of the femur, and the hip joint. The medial circumflex femoral artery is the main source of arterial blood supply to the head and neck of the femur.

  (2) The profunda femoral artery gives rise to 4 perforating arteries, which supply the medial thigh and pass through the adductor magnus to supply the muscles in the posterior thigh.

THE CRUCIATE ANASTOMOSIS

*The medial and lateral circumflex femoral arteries, the inferior gluteal artery, and the first perforating artery contribute to the **cruciate anastomosis in the posterior thigh**. The cruciate anastomosis may contribute to collateral circulation of the lower limb if the femoral artery becomes occluded.*

 2. The **popliteal artery**

  a. The popliteal artery begins at the adductor hiatus as a continuation of the femoral artery (see Figure 7–4b).

  b. It courses through the popliteal fossa posterior to the knee with the tibial nerve.

  c. The popliteal artery gives rise to 5 genicular arteries, which supply the knee joint.

  d. The 5 genicular branches of the popliteal artery, the descending genicular branch of the femoral artery, the descending branch of the lateral circumflex femoral artery, and the anterior recurrent branch of the anterior tibial artery contribute to **collateral circulation around the knee.**

  e. The popliteal artery ends at the inferior border of the popliteus by dividing into anterior and posterior tibial arteries (see Figure 7–4b).

 3. The **anterior tibial artery**

  a. The anterior tibial artery enters the anterior compartment of the leg proximal to the interosseous membrane between the tibia and the fibula.

Table 7–5. Muscles that act at joints of the toes.

Action	Muscles Involved	Inneration	Major Segments
Flexion			
Great toe (DIP joint)	Flexor hallucis longus	Tibial	L5, S1
Great toe (DIP joint)	Flexor hallucis longus	Medial plantar	L5, S1
Toes 2–5 (DIP joint)	Flexor digitorum longus (FD)	Tibial	L5, S1
Straightens tendon of FD longus	Quadratus plantae (Fl. accessorius)	Lateral plantar	S1, 2
Toes 2–5 (PIP joints)	Flexor digitorum brevis	Medial plantar	L5, S1
Toe 5	Flexor digiti minimi brevis	Lateral plantar	S1, 2
Toes 2–5 (MP joints)	Lumbrical I Lumbricals II, III, IV Interossei	Medial plantar, toe 2 Lateral plantar, toes 3–5 Lateral plantar	L5, S1 S1, 2 S1, 2
Extension			
Great toe	Extensor hallucis longus Extensor hallucis brevis	Deep peroneal Deep peroneal	L5, S1 L5, S1
Toes 2–4	Extensor digitorum longus Extensor digitorum brevis	Deep peroneal Deep peroneal	L4, 5, S1 L5, S1
Toes 2–5 (IP joints)	Lumbricals	Lateral plantar, toes 3–5 Medial plantar, toe 2	L5, S1 S1, 2
Abduction at MP (to/from midline of toe 2)			
Toe 1	Abductor hallucis	Medial plantar	L5, S1
Toes 2–4	Dorsal interossei	Lateral plantar	S1, 2
Toe 5	Abductor digiti minimi	Lateral plantar	S1, 2
Adduction at TMP (to midline of toe 2)			
Toe 1	Adductor hallucis	Lateral plantar	S1, 2
Toes 3–5	Plantar interossei	Lateral plantar	S1, 2

DIP, distal interphalangeal joints; PIP, proximal interphalangeal joints; MP metatarsophalangeal joints.

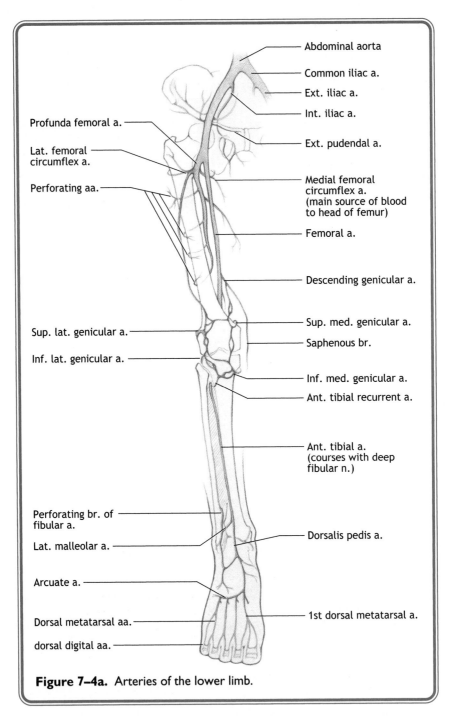

- Abdominal aorta
- Common iliac a.
- Ext. iliac a.
- Int. iliac a.
- Ext. pudendal a.

Profunda femoral a. —

Lat. femoral circumflex a. —

Perforating aa. —

- Medial femoral circumflex a. (main source of blood to head of femur)
- Femoral a.
- Descending genicular a.
- Sup. med. genicular a.
- Saphenous br.

Sup. lat. genicular a. —

Inf. lat. genicular a. —

- Inf. med. genicular a.
- Ant. tibial recurrent a.
- Ant. tibial a. (courses with deep fibular n.)

Perforating br. of fibular a. —

Lat. malleolar a. —

- Dorsalis pedis a.

Arcuate a. —

Dorsal metatarsal aa. —

dorsal digital aa. —

- 1st dorsal metatarsal a.

Figure 7–4a. Arteries of the lower limb.

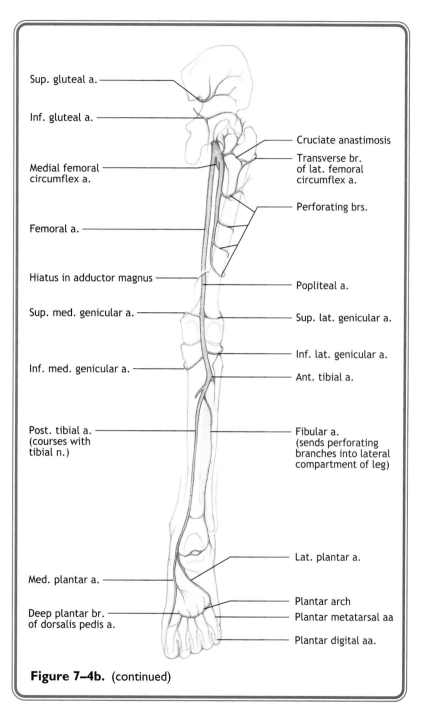

Sup. gluteal a.

Inf. gluteal a.

Cruciate anastimosis

Transverse br.
of lat. femoral
circumflex a.

Medial femoral
circumflex a.

Perforating brs.

Femoral a.

Hiatus in adductor magnus

Popliteal a.

Sup. med. genicular a.

Sup. lat. genicular a.

Inf. lat. genicular a.

Inf. med. genicular a.

Ant. tibial a.

Post. tibial a.
(courses with
tibial n.)

Fibular a.
(sends perforating
branches into lateral
compartment of leg)

Lat. plantar a.

Med. plantar a.

Plantar arch

Deep plantar br.
of dorsalis pedis a.

Plantar metatarsal aa

Plantar digital aa.

Figure 7–4b. (continued)

b. The anterior tibial artery courses with the deep fibular (peroneal) nerve and supplies the anterior compartment of the leg.

    c. It continues as the dorsalis pedis, on the dorsal aspect of the foot. The dorsalis pedis branches into an arcuate artery, which gives rise to digital branches that supply the toes, and a deep plantar artery, which contributes to a plantar arterial arch in the sole of the foot.

DORSALIS PEDIS PULSE

*A **dorsalis pedis pulse** may be evaluated by compressing the dorsal artery of the foot against the tarsal bones lateral to the tendon of the extensor hallucis longus.*

    4. The **posterior tibial artery**

        a. The posterior tibial artery arises from the popliteal artery and courses through the posterior compartment of the leg with the tibial nerve.

        b. It supplies the posterior compartment of the leg.

        c. The posterior tibial artery gives rise to the fibular artery, which supplies the posterior compartment of the leg and sends perforating branches into the lateral compartment of the leg to supply the fibularis longus and brevis muscles.

        d. It passes into the foot behind the medial malleolus adjacent to the tendons of the tibialis posterior and flexor digitorum longus muscles and divides into the medial and lateral plantar arteries, which supply the sole of the foot.

B. **Branches of the internal iliac artery** supply the gluteal region and medial thigh and include the superior gluteal, inferior gluteal, and obturator arteries (see Figure 7–4b).

    1. The **superior gluteal artery**

        a. The superior gluteal artery enters the gluteal region with the superior gluteal nerve superior to the piriformis after passing through the greater sciatic foramen.

        b. It supplies the gluteus maximus, medius, and minimus muscles.

    2. The **inferior gluteal artery**

        a. The inferior gluteal artery enters the gluteal region with the inferior gluteal nerve inferior to the piriformis muscle after passing through the greater sciatic foramen.

        b. It supplies the gluteus maximus muscle, short lateral rotators of the hip, and proximal parts of the hamstrings.

    3. The **obturator artery**

        a. The obturator artery enters the medial thigh with the obturator nerve after passing through the obturator foramen.

        b. It supplies the adductor muscles, obturator externus, pectineus, and gracilis muscles.

III. Venous Drainage

A. The **lower limb is drained** by a superficial and a deep system of veins.

B. The **great saphenous and small saphenous veins** are the superficial veins.

    1. The great saphenous vein

        a. The great saphenous vein arises from the medial aspect of the dorsal venous arch of the foot.

b. It courses anterior to the medial malleolus, through the medial aspect of the leg with the saphenous nerve, and through the medial thigh.

      c. The great saphenous vein drains into the femoral vein after passing through the saphenous hiatus, a fault in the fascia lata.

  2. The small saphenous vein

      a. The small saphenous vein arises from the lateral aspect of the dorsal venous arch of the foot.

      b. It courses posterior to the lateral malleolus and then through the posterior leg with the sural nerve.

      c. The small saphenous vein passes between the 2 heads of the gastrocnemius muscle and drains into the popliteal vein.

  3. The **deep veins course with arteries of the same name**; perforating veins connect the deep veins with the superficial veins.

IV. Innervation

  A. Nerves that arise from the **lumbar and the lumbosacral plexuses** innervate the lower limb. The lumbar plexus is formed by the ventral rami of L1 through L4 with a small contribution from T12 (Figure 7–5) and is found on the posterior abdominal wall and greater pelvis.

    1. The **ventral rami of the lumbar plexus** branch into posterior and anterior divisions.

    2. The **posterior and anterior divisions** form 2 main nerves: the femoral and obturator nerves, respectively.

  B. The **lumbosacral plexus** is formed by the ventral rami of L4 through S3 (Figure 7–6) and is found in the lesser pelvis.

    1. The **ventral rami** of the lumbosacral plexus branch into posterior and anterior divisions.

    2. The **posterior and anterior divisions** form 2 terminal nerves: the common fibular (peroneal) and tibial nerves, respectively. The superior and inferior gluteal nerves also contain posterior division fibers.

SIGNIFICANCE OF ANTERIOR AND POSTERIOR DIVISION FIBERS

*During development, the **lower limb** undergoes a **medial rotation** so that the flexor muscles that were anterior in the embryo come to be situated posteromedially and extensor muscles that were posterior in the embryo come to be situated anterolaterally. As a result, the medial and posterior compartments of the thigh, the posterior compartment of the leg, and the plantar muscles of the foot are innervated by the obturator or tibial nerves, which contain anterior division fibers.*

Muscles in the anterior compartment of the thigh and the anterior and lateral compartments of the leg and the dorsum of the foot are innervated by the femoral or common fibular nerves, which contain posterior division fibers.

  C. In the plexus, **fibers of L4 unite with fibers of L5** to form the lumbosacral trunk.

    1. The lumbosacral trunk emerges from the medial aspect of the psoas major muscle, crosses the pelvic brim, and joins with the ventral rami of the first, second, and third sacral nerves to form the lumbosacral plexus.

    2. The **terminal and collateral nerves** of the lumbosacral plexus exit the pelvis through the greater sciatic foramen.

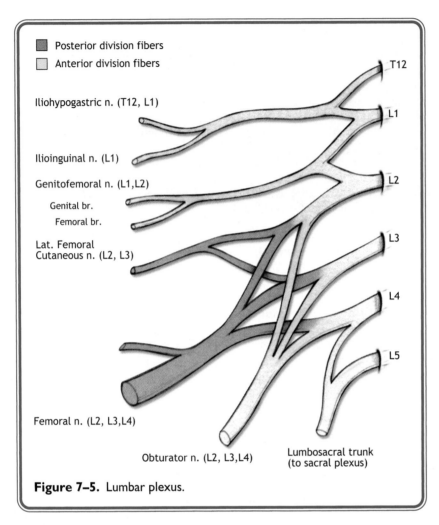

Posterior division fibers
Anterior division fibers

Iliohypogastric n. (T12, L1)

Ilioinguinal n. (L1)

Genitofemoral n. (L1,L2)

  Genital br.
  Femoral br.

Lat. Femoral
Cutaneous n. (L2, L3)

Femoral n. (L2, L3,L4)

Obturator n. (L2, L3,L4)

Lumbosacral trunk
(to sacral plexus)

T12
L1
L2
L3
L4
L5

Figure 7–5. Lumbar plexus.

GRADIENT OF INNERVATION

*The **ventral rami** of the lumbar and lumbosacral plexus exhibit a proximal to distal gradient of innervation.*
- *Nerves that contain fibers from the superior rami of the plexus (L2 through L4) innervate muscles in the anterior and medial thigh that act at the hip and knee joints.*
- *Nerves that contain fibers from the inferior rami of the plexus (S1 through S3) innervate muscles of the leg that act at the joints of the ankle and foot.*

CLINICAL
CORRELATION

V. **Terminal Nerves of the Lumbar Plexus**

  A. **Femoral Nerve (Figures 7–5 and 7–8)**
    1. The **femoral nerve** contains posterior division fibers from L2, L3, and L4 ventral rami.
    2. It emerges from the lateral border of the psoas major in the iliac fossa and passes into the anterior thigh posterior to the inguinal ligament and lateral to the femoral artery.

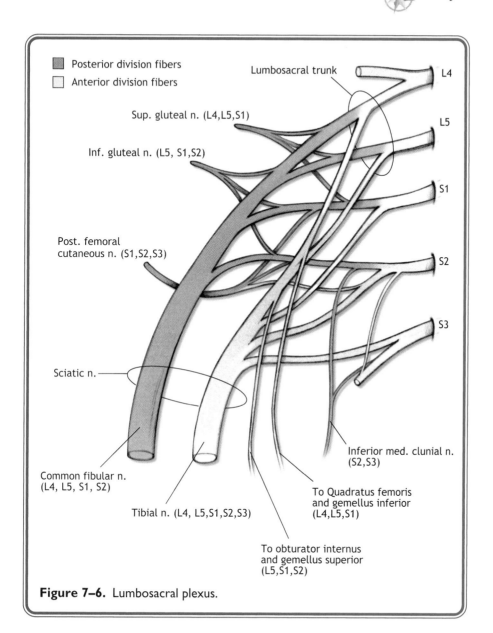

□ Posterior division fibers
□ Anterior division fibers

Lumbosacral trunk

Sup. gluteal n. (L4,L5,S1)

Inf. gluteal n. (L5, S1,S2)

Post. femoral
cutaneous n. (S1,S2,S3)

Sciatic n.

Common fibular n.
(L4, L5, S1, S2)

Tibial n. (L4, L5,S1,S2,S3)

To obturator internus
and gemellus superior
(L5,S1,S2)

To Quadratus femoris
and gemellus inferior
(L4,L5,S1)

Inferior med. clunial n.
(S2,S3)

L4

L5

S1

S2

S3

Figure 7–6. Lumbosacral plexus.

3. The femoral nerve innervates muscles in the anterior compartment of the thigh, including the 4 heads of the quadriceps femoris, iliopsoas, sartorius, and pectineus, which collectively act to flex the thigh at the hip and extend the leg at the knee.

4. It innervates skin of the anterior and medial thigh (medial and intermediate cutaneous nerves).

5. The femoral nerve gives rise to the saphenous nerve (L3, L4) (Figure 7–8).

   a. The **saphenous nerve** is the **longest branch of the femoral** nerve and is the only branch of the lumbar plexus to cross the knee joint.

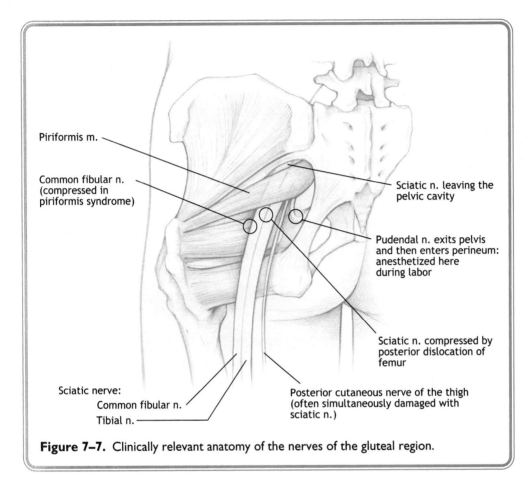

Figure 7–7. Clinically relevant anatomy of the nerves of the gluteal region.

b. The saphenous nerve enters the adductor canal but leaves the canal without passing through the adductor hiatus.

c. The saphenous nerve courses with the great saphenous vein and innervates skin of the medial side of the leg and foot.

FEMORAL NERVE LESIONS (TABLE 7–6)

*The **femoral nerve** may be damaged in the abdomen by an abscess of the psoas major. Patients experience weakness in the ability to flex the thigh at the hip, a weakness in the ability to extend the leg at the knee, and a diminished patellar tendon reflex.*

SAPHENOUS NERVE LESIONS

*The **saphenous nerve** may be lesioned during a surgical procedure of the leg to remove part of the great saphenous vein, or it may be lacerated as it pierces the wall of the adductor canal. Patients experience pain and paresthesia in the skin of the medial aspect of the leg and foot.*

B. **Obturator Nerve** (see Figure 7–5)

1. The **obturator nerve** contains anterior division fibers from the L2, L3, and L4 ventral rami.

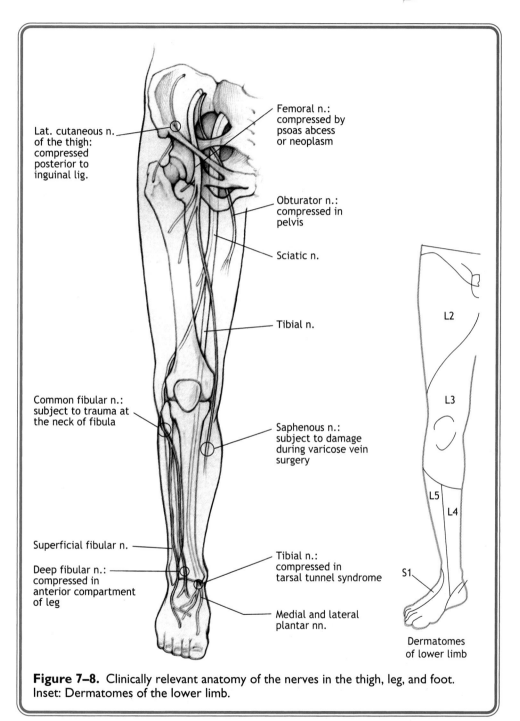

Lat. cutaneous n. of the thigh: compressed posterior to inguinal lig.

Femoral n.: compressed by psoas abcess or neoplasm

Obturator n.: compressed in pelvis

Sciatic n.

Tibial n.

Common fibular n.: subject to trauma at the neck of fibula

Saphenous n.: subject to damage during varicose vein surgery

Superficial fibular n.

Deep fibular n.: compressed in anterior compartment of leg

Tibial n.: compressed in tarsal tunnel syndrome

Medial and lateral plantar nn.

L2

L3

L5

L4

S1

Dermatomes of lower limb

Figure 7–8. Clinically relevant anatomy of the nerves in the thigh, leg, and foot. Inset: Dermatomes of the lower limb.

Table 7–6. Effects of lesions to major roots and nerves of lumbosacral plexus.

Lesioned Root	Cause of Lesion	Location of Dermatome	Muscles Affected	Suppressed Reflex
L4	Osteoarthritis	Medial leg	Quadriceps, iliopsoas; Hip adductors	Patellar tendon
L5	Herniation of disk between L4 and L5	Dorsum of foot; Great toe, toes 2 and 3	Foot dorsiflexors; Toe extensors	
S1	Herniation of disk between L5 and S1	Posterior leg, lateral foot, toes 4, 5, and sole	Plantar flexors	Achilles tendon

Lesioned Nerve	Cause of Lesion	Altered Cutaneous Sensation	Weakness in	Common Sign of Lesion
Obturator (L2, L3, L4)	Pelvic neoplasm; Pregnancy	Medial thigh	Adduction at hip	
Femoral (L2, L3, L4)	Diabetes; Pelvic neoplasm; Psoas abscess	Anterior thigh, medial leg to medial malleolus	Flexion of hip; Extension of leg at knee	
Common fibular (L4, L5, S1, S2)	Compression at neck of fibula; Hip fracture; Dislocation of femur; Piriformis syndrome	Anterior and lateral leg, dorsum of foot	Dorsiflexion, eversion of foot	Foot drop; Steppage gait
Tibial (L4, L5, S1, S2, S3)	Diabetes; Hip fracture; Dislocation of femur	Posterior leg, sole and lateral foot	Plantar flexion; Flexion of toes	Can't stand on "tiptoes"
Superior gluteal (L4, L5, S1)	Misplaced gluteal injection; Pelvic neoplasm		Abduction at hip	Pelvic tilt; Waddling gait
Inferior gluteal (L5, S1, S2)	Pelvic neoplasm		Extension at hip from flexed position	Can't get up from chair

2. It emerges from the medial side of the psoas major, crosses the pelvic brim, and courses anteriorly and inferiorly in the lesser pelvis to the obturator foramen.

3. The obturator nerve passes through the obturator foramen and through the obturator externus into the medial thigh.

4. It divides into an anterior branch, which passes between the adductor longus and brevis muscles, and a posterior branch, which passes between the adductor brevis and adductor magnus.

5. The obturator nerve innervates muscles in the medial thigh, including the adductor longus, brevis and magnus, gracilis, and obturator externus muscles, which act mainly to adduct the thigh at the hip and assist in flexing the thigh at the hip.

6. The obturator nerve innervates skin in a small region of the medial thigh.

OBTURATOR NERVE LESIONS (TABLE 7-6)

*The **obturator nerve** is most commonly lesioned in the pelvis. Patients are unable to adduct the thigh at the hip and may have paresthesia in skin of the medial thigh (Figure 7-8).*

VI. Five Collateral Nerves of the Lumbar Plexus (see Figure 7-5)

A. Subcostal Nerve (T12)

1. The **subcostal nerve** passes between the psoas major and quadratus lumborum muscles inferior to the 12th rib.

2. It innervates abdominal musculature and overlying skin of the lateral and anterior abdominal wall.

B. Iliohypogastric Nerve (T12-L1)

1. The **iliohypogastric nerve** emerges between the psoas major and quadratus lumborum muscles inferior to the subcostal nerve.

2. It innervates abdominal musculature and the skin of the inguinal and hypogastric regions of the lateral and anterior abdominal wall.

C. Ilioinguinal Nerve (L1)

1. The **ilioinguinal nerve** courses inferior to the iliohypogastric nerve.

2. It innervates abdominal musculature and the skin of the inguinal and hypogastric regions of the lateral and anterior abdominal wall.

3. The ilioinguinal nerve pierces the inguinal canal and passes through the superficial inguinal ring to innervate the skin of the medial thigh, labium majus, and anterior aspect of the scrotum.

D. Genitofemoral Nerve (L1, L2)

1. The **genitofemoral nerve** courses through and then anterior to the psoas major muscle.

2. It divides into a femoral branch and a genital branch.

   a. The femoral branch passes posterior to the inguinal ligament and innervates the skin of the medial thigh.

   b. The genital branch enters the inguinal canal through the deep inguinal ring and innervates the cremasteric muscle.

E. Lateral Femoral Cutaneous Nerve (L2, L3)

1. The **lateral femoral cutaneous nerve** emerges lateral to the psoas major muscle and then crosses the iliacus to reach the anterior superior iliac spine.

2. It descends into the lateral thigh after passing posterior to the inguinal ligament.

LATERAL FEMORAL CUTANEOUS NERVE LESIONS

*The **lateral femoral cutaneous nerve** may be compressed as it passes posterior to the lateral part of the inguinal ligament just medial to the anterosuperior iliac spine. Patients with compression of the lateral femoral cutaneous nerve (meralgia paresthetica) present with pain and paresthesia in the anterolateral thigh.*

VII. Terminal Nerves of the Lumbosacral Plexus

   A. Superior Gluteal Nerve (Posterior Division Fibers from L4, L5, and S1) (Figure 7–6)

      1. The **superior gluteal nerve** enters the gluteal region with the superior gluteal artery by passing through the greater sciatic foramen superior to the piriformis muscle.

      2. It innervates the gluteus medius, gluteus minimus, and the tensor fasciae latae muscles.

SUPERIOR GLUTEAL NERVE LESIONS

*Patients with a **lesion of the superior gluteal** nerve have a weakness in the ability to abduct the thigh at the hip.*
- *Patients experience a **waddling or Trendelenburg gait**, in which the pelvis sags on the side of the unsupported limb.*
- *The pelvis sags on the side that is opposite the side of the lesioned superior gluteal nerve.*

   B. Inferior Gluteal Nerve (Posterior Division Fibers from L5, S1, and S2) (Figure 7–6)

      1. The **inferior gluteal nerve** enters the gluteal region by passing through the greater sciatic foramen inferior to the piriformis muscle.

      2. It innervates the gluteus maximus muscle.

INFERIOR GLUTEAL NERVE LESIONS

*Patients with a **lesion of the inferior gluteal nerve** have a weakness in the ability to laterally rotate and extend the thigh at the hip.*
- *Patients have **difficulty extending the thigh at the hip from a flexed position**, as in climbing stairs or rising from a chair.*
- *Patients may have a **gluteus maximus gait**, in which they thrust their torso posteriorly in an attempt to counteract the weakness of the gluteus maximus.*

   C. Tibial Nerve (Anterior Division Fibers from L4, L5, S1, S2, and S3) (Figures 7–6 to 7–9)

      1. The **tibial nerve** enters the gluteal region with the common fibular nerve in the sciatic nerve by passing through the greater sciatic foramen inferior to the piriformis muscle.

      2. It courses through the posterior thigh deep to the hamstrings before separating from the common fibular nerve at the superior border of the popliteal fossa.

      3. The tibial nerve courses in the posterior part of the leg with the posterior tibial artery and then passes through the tarsal tunnel and into the sole of the foot after coursing behind the medial malleolus.

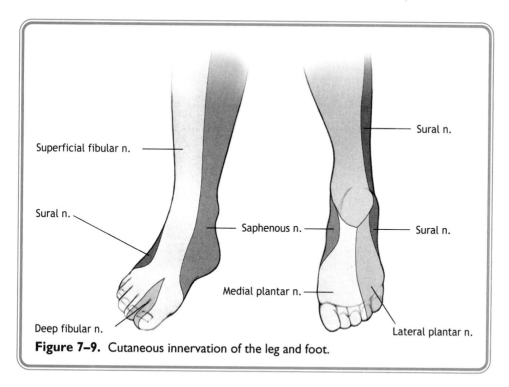

Figure 7–9. Cutaneous innervation of the leg and foot.

4. The tibial nerve innervates muscles in the posterior thigh, posterior leg, and plantar foot.
5. It divides into the medial and lateral plantar nerves distal to the tarsal tunnel.
   a. The distribution of the **medial plantar nerve** (L5, S1, S2) is similar to that of the median nerve in the hand.
   b. The medial plantar nerve innervates 4 muscles in the sole of the foot: **flexor digitorum brevis, flexor hallucis brevis, abductor hallucis, and the first lumbrical**.
   c. Common and digital plantar branches of the medial plantar nerve innervate the skin of the medial side of the sole of the foot and the medial 3 1/2 digits.
   d. Distribution of the **lateral plantar nerve** (L5, S1, S2) is similar to that of the ulnar nerve of the hand.
   e. The lateral plantar nerve innervates the rest of the intrinsic foot muscles.
   f. Common and digital plantar branches of the lateral plantar nerve innervate skin of the lateral side of the sole of the foot and the lateral 1 1/2 digits.

SCIATIC NERVE LESIONS (TABLE 7–6)

The **sciatic nerve** is susceptible to damage from an intramuscular injection in the lower medial quadrant of the gluteus maximus muscle, or it may be compressed as a result of a posterior dislocation of the femur. Such injuries may affect both the common fibular and tibial nerves (see next discussion). The L5 and S1 roots of the sciatic nerve are commonly compressed by intervertebral disk herniations. Patients have pain that radiates into the L5 and S1 dermatomes of the leg and foot (Figure 7–8 inset).

TIBIAL NERVE LESIONS

*In patients with **tibial nerve lesions** in the gluteal region, weakness may be evident in the ability to flex the leg at the knee and plantar flex at the ankle.*

*The **tibial nerve** may be compressed at the ankle as it courses through the tarsal tunnel adjacent to the medial malleolus. Patients with tarsal tunnel syndrome have pain and paresthesia in the sole of the foot Figure 7–8).*

    D. Common Fibular (Peroneal) Nerve (Posterior Division Fibers from L4, L5, S1, and S2) (Figures 7–6 to 7–9)
        1. The **common fibular nerve** emerges from the pelvis with the tibial nerve inferior to the piriformis muscle. It courses with the tibial nerve into the posterior thigh.
        2. The common fibular nerve innervates the short head of the biceps femoris muscle in the posterior thigh and then separates from the tibial at the superior border of the popliteal fossa.
        3. It courses laterally along the tendon of the biceps femoris muscle and spirals around the neck of the fibula into the lateral compartment of the leg.
        4. The common fibular nerve enters the fibularis longus muscle and divides into the superficial fibular and deep fibular nerves.

COMMON FIBULAR NERVE LESIONS (TABLE 7–6)

*The **common fibular nerve** is the most frequently lesioned nerve in the lower limb.*
- *The common fibular nerve is frequently lesioned as it passes around the neck of the fibula.*
- *Patients experience **footdrop**, which results from a loss of dorsiflexion at the ankle, and a loss of eversion. Patients have pain and paresthesia in the lateral leg and dorsum of the foot.*
- *Patients with footdrop may have a **steppage gait**, in which they raise their affected leg high off the ground and their foot slaps the ground when walking.*
- *In **piriformis syndrome**, the common fibular nerve may be compressed by the fibers of the piriformis muscle when the nerve passes through the piriformis rather than anterior to it with the tibial nerve (see Figure 7–7).*

    E. Superficial Fibular Nerve (L4, L5, and S1) (Figures 7–8 and 7–9)
        1. The **superficial fibular nerve** innervates the fibularis longus and brevis muscles in the lateral compartment of the leg.
        2. It emerges in the distal third of the lateral leg and innervates skin of the lateral leg and dorsum of the foot except for the first dorsal webbed space between the great toe and the second toe.

SUPERFICIAL FIBULAR NERVE LESIONS

*The **superficial fibular nerve** may be lesioned as the nerve emerges from the lateral compartment of the leg. Patients experience pain and paresthesia in the dorsal aspect of the foot.*

    F. Deep Fibular Nerve (L5, S1, and S2) (Figure 7–8)
        1. The **deep fibular nerve** courses through the fibularis longus muscle and the anterior compartment of the leg with the anterior tibial artery.
        2. It innervates muscles in the anterior compartment of the leg, including the tibialis anterior, extensor hallucis longus, extensor digitorum longus, and peroneus tertius muscles.
        3. The deep fibular nerve innervates muscles in the dorsum of the foot, including the extensor digitorum brevis and extensor hallucis brevis muscles.
        4. It innervates skin of the webbed space between the great toe and the second toe.

DEEP FIBULAR NERVE LESIONS

*The **deep fibular nerve** may be compressed in the anterior compartment of the leg. These patients may have footdrop and paresthesia in skin of the webbed space between the great toe and the second toe.*

VIII. Collateral Nerves of the Lumbosacral Plexus (see Figure 7–6)

A. Sural Nerve
1. The **sural nerve** innervates the skin of the posterior leg and lateral aspect of the foot.
2. It receives contributions from the tibial and common fibular nerves.
3. The sural nerve courses with the small saphenous vein in the posterior leg.

B. Nerve to the Quadratus Femoris Muscle (L4, L5, and S1)
1. The **nerve to the quadratus femoris muscle** enters the gluteal region after passing through the greater sciatic foramen.
2. It innervates the inferior gemellus and quadratus femoris muscles.

C. Nerve to the Obturator Internus Muscle (L5, S1, and S2)
1. The **nerve to the obturator internus muscle** enters the gluteal region after passing through the greater sciatic foramen and innervates the superior gemellus.
2. It crosses the ischial spine and passes through the lesser sciatic foramen to innervate the obturator internus.

D. Posterior Femoral Cutaneous Nerve (Anterior and Posterior Division Fibers from S1, S2, and S3)
1. The **posterior femoral cutaneous nerve** innervates the skin of the posterior thigh and upper calf.
2. It is the only branch of the lumbosacral plexus that contains both anterior division (S2, S3) and posterior division (S1, S2) fibers.

E. Nerve to the Piriformis Muscle (S1, S2)
1. The **nerve to the piriformis muscle** arises from the first and second sacral ventral rami.
2. It innervates the piriformis muscle.

F. Perforating Cutaneous Nerve (S2, S3)
1. The **perforating cutaneous nerve** arises from the second and third sacral ventral rami.
2. It innervates skin covering the ischioanal fossa and the gluteal region near the anal canal.

CLINICAL PROBLEMS

Match the deficit in Questions 1–10 with the appropriate injury in Choices A–H. (Choices may be used once, more than once, or not at all.)

  A. Femoral nerve

  B. Tibial nerve

  C. Common fibular (peroneal) nerve

  D. Deep fibular (peroneal) nerve

E. Superficial fibular (peroneal) nerve

F. Superior gluteal nerve

G. Inferior gluteal nerve

H. Obturator nerve

1. Patient has a weakness in the ability to dorsiflex the foot and has altered sensation in skin between the great toe and the second toe.

2. Patient has altered sensation in skin covering the great saphenous vein in the leg.

3. Patient has weakness in the ability to abduct the thigh at the hip.

4. Patient has weakness of muscles in the ability to evert the foot and has altered sensation in skin of the dorsal aspect of the foot; the ability to extend the toes is intact.

5. Patient has weakness in the ability to extend the thigh at the hip when arising from a chair, and difficulty walking up stairs; cutaneous sensation is intact.

6. Patient develops tarsal tunnel syndrome and has weakness in the ability to flex the great toe.

7. Patient is a ballet dancer who can no longer stand on her "tip toes."

8. When a limb is off the ground during gait, patient experiences lateral pelvic tilt toward the unsupported side.

9. Patient has weakness in the ability to adduct the thigh at the hip.

10. Patient has weakness in the ability to flex the thigh at the hip and extend the leg at the knee.

A football player suffers trauma to the lateral part of the leg just distal to the head of the fibula and a nerve is lesioned.

11. What might the patient experience?

A. Weakness in the ability to plantar flex the foot

B. Loss of the ability to invert the foot

C. Altered sensation in the skin of the medial aspect of the leg

D. Altered sensation in the skin of the dorsal aspect of the foot

E. Weakened ability to flex the toes

An elderly woman who suffers from osteoporosis falls and "breaks her hip." The orthopedic surgeon recommends that the proximal part of the femur be replaced with a prosthesis because of the likelihood of avascular necrosis of the head of the femur.

12. What artery supplying the neck and head of the femur might have been lacerated by the fracture?

A. Inferior gluteal artery

B. Medial circumflex femoral artery

C. Pudendal artery

D. Profunda femoral artery

E. Obturator artery

A traumatic injury to the lateral aspect of a patient's knee tears several structures at the knee joint. An examination reveals a positive anterior drawer sign and a clicking sound when the patient attempts to extend the leg at the knee.

13. Of the following structures, which one was most likely spared from being stretched or torn in this knee injury?

A. Medial meniscus

B. Tibial collateral ligament

C. Fibular collateral ligament

D. Anterior cruciate ligament

E. Tendon of the sartorius muscle

A health care worker inadvertently administers an injection to the gluteal region that results in a lesion to a nerve. The patient begins to walk with an altered gait. Upon raising the left foot off the ground during gait, the patient leans to the right, and when standing on the right foot without leaning, the left buttock seems to sag.

14. What muscle might have been weakened by the nerve lesion?

A. Gluteus maximus muscle

B. Gluteus medius muscle

C. Piriformis muscle

D. Semitendinosus muscle

E. Quadratus femoris muscle

A college cross-country runner begins to experience leg and foot pain during and after training for the upcoming season. The pain radiates from the anterolateral leg into the dorsal aspect of the foot. Dorsiflexion and extension of the toes are performed only with pain. The leg appears to be swollen in the area of the pain.

15. Which of the following arteries may have been compressed by the swelling?

A. Popliteal artery

B. Anterior tibial artery

C. Posterior tibial artery

D. Peroneal artery

E. Medial plantar artery

A 56-year-old man develops numbness and tingling in the lower limb followed by progressive muscular weakness. You suspect that the man's peripheral neuropathy may be a side effect of a drug that he is taking. You order a biopsy of a cutaneous nerve in the posterior leg that accompanies the small saphenous vein.

16. Which of the following nerves was correctly biopsied?

A. Tibial

B. Superficial fibular

C. Sural

D. Saphenous

E. Deep fibular

ANSWERS

1. D

2. A

3. F

4. E

5. G

6. B

7. B

8. F

9. H

10. A

11. The answer is D. Lesions of the common fibular nerve result in footdrop and altered sensation in the skin of the dorsal aspect of the foot.

12. The answer is B. The head of the femur may undergo avascular necrosis as a result of the disruption of branches of the medial circumflex femoral artery, the main source of arterial blood supply to the head and neck of the femur.

13. The answer is C. A traumatic blow to the lateral aspect of the knee most commonly injures the tibial collateral ligament, the medial meniscus, and the anterior cruciate ligament (the terrible triad) and may tear tendons that cross the medial aspect of the joint.

14. The answer is B. A lesion of the superior gluteal nerve may result in weakness in the ability to abduct the thigh at the hip and keep the pelvis level during gait. Patients have a waddling, or Trendelenburg, gait, in which the pelvis sags on the side of the unsupported limb because of weakness of the gluteus medius muscle.

15. The answer is B. The anterior tibial artery courses through the anterior compartment and may be compressed by the swelling.

16. The answer is C. The sural nerve is a cutaneous nerve that innervates the posterior leg. The saphenous nerve supplies the medial leg; the superficial fibular nerve supplies the lateral leg. The tibial and deep fibular nerves do not have cutaneous branches in the leg.

CHAPTER 8
HEAD

I. The **skull** consists of the cranium and the facial skeleton.

   A. The **cranium** consists of 8 bones, which form the calvaria and the cranial base (Figure 8–1).

      1. Bones of the calvaria consist of 2 layers of compact bone separated by the **diploë**, a layer of bone marrow.

      2. The **bones of the cranial base** articulate with the atlas, the bones of the facial skeleton, and the mandible.

      3. The **bones of the cranium** are joined by sutures.

         a. The **coronal suture** is formed by the frontal bones and the 2 parietal bones.

         b. The **sagittal suture** is formed by the 2 parietal bones.

         c. The **lambdoidal suture** is formed by the 2 parietal bones and the occipital bone.

      4. The **pterion** is on the lateral aspect of the skull superior to the zygomatic arch and posterior to the lateral wall of the orbit. The pterion is superficial to the anterior branch of the middle meningeal artery, which supplies the dura and the skull.

SKULL FRACTURE AT THE PTERION

*A **lateral skull fracture at the pterion**, the thinnest part of the calvaria, may lacerate the middle meningeal artery and cause an **epidural or extradural hematoma**.*

- *The epidural arterial hemorrhage forms a biconvex lens-shaped hematoma between the skull and the periosteal dura, which does not pass the sutures.*
- *An epidural hematoma may compress the lateral part of a cerebral hemisphere and result in herniation of the medial part of the temporal lobe through the tentorial notch of the dura. The herniated temporal lobe may compress the brainstem.*
- *The patient may have an initial lucid asymptomatic interval, followed by weakness of limb muscles, a dilated pupil resulting from compression of the oculomotor nerve (CN III), and deterioration of cardiovascular and respiratory functions.*

CLINICAL CORRELATION

   B. The **facial skeleton** is formed by 14 bones, which enclose the orbits, nasal cavity, oral cavity, and paranasal sinuses.

II. The **cranial cavity** contains the brain and the cranial nerves.

   A. The **internal carotid and vertebral arteries** and their branches supply the brain; branches of these arteries anastomose to form the circle of Willis (Figure 8–2).

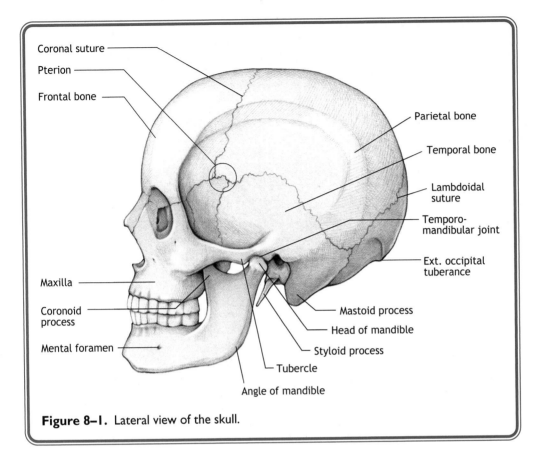

Figure 8–1. Lateral view of the skull.

1. The vertebral arteries enter the skull through the **foramen magnum** and unite to form the basilar artery; the basilar artery divides into a pair of posterior cerebral arteries.
2. Each internal carotid artery enters the skull through the **carotid canal**, crosses the width of the foramen lacerum, and passes through the cavernous sinus. At the base of the brain, each internal carotid artery divides into a small anterior cerebral artery and a large middle cerebral artery.
3. The **circle of Willis** is completed by an anterior communicating artery, which connects the anterior cerebral arteries, and a pair of posterior communicating arteries, which connect the internal carotid arteries with the posterior cerebral arteries.

BERRY ANEURYSMS

*Berry aneurysms (saccular dilatations of the walls of arteries) most commonly occur in the anterior part of the circle of Willis at branch points of the anterior communicating artery, posterior communicating artery, or middle cerebral artery. Blood from a ruptured aneurysm may accumulate in the subarachnoid space and cause a **subarachnoid hematoma**.*

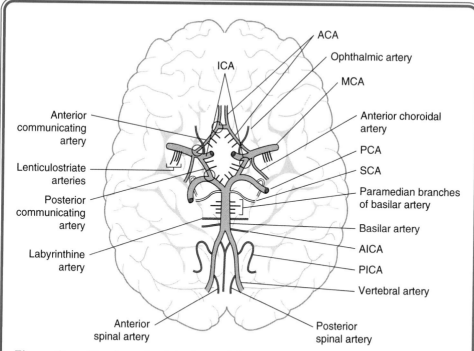

Figure 8–2. Blood supply of the brain. ACA, anterior cerebral artery; AICA, anterior inferior cerebellar artery; ICA, internal carotid artery; MCA, middle cerebral artery; PCA, posterior cerebral artery; PICA, posterior inferior cerebellar artery; SCA, superior cerebellar artery. Circled areas indicate common sites of berry aneurysms.

- *Patients with a **ruptured berry aneurysm** may experience an acute explosive "worst headache of my life." The **headache** is caused by blood leaking from the aneurysm, which irritates the meninges. Patients also may have a stiff neck resulting from irritation of spinal dura.*

- *The **oculomotor nerve** (CN III) may be compressed by an aneurysm at the junction of the posterior communicating artery and internal carotid artery or the posterior cerebral artery. Patients with **compression of the oculomotor nerve** initially have a **dilated or a "blown" pupil**.*

   B. The **meninges of the cranial cavity** consist of 2 layers of dura mater, the arachnoid mater, and the pia mater.

      1. The **dura** consists of an outer periosteal layer and an inner meningeal layer.

         a. The **outer periosteal layer** is closely applied to the inner surface of bones of the skull and is firmly attached to sutures.

         b. The **inner meningeal layer** is continuous with dura of the vertebral canal.

            (1) The meningeal dura forms septa that extend between the 2 cerebral hemispheres (falx cerebri), between the hemispheres and the cerebellum (tentorium cerebelli), and cover the pituitary gland in the sella turcica (diaphragma sellae).

(2) Most of the meningeal dura is innervated by meningeal branches of the ophthalmic, maxillary, and mandibular divisions of the trigeminal nerve.

(3) The meningeal dura of the posterior cranial fossa is innervated by the first 3 cervical spinal nerves, and by meningeal branches the vagus nerve. (CN X).

HEADACHE

*The **meningeal dura** is sensitive to pain; irritation or stretching of the dura is a common cause of **headache**. Pain is commonly referred to regions supplied by branches of the trigeminal nerve.*

2. The arachnoid mater is closely applied to the dura; the subarachnoid space containing cerebrospinal fluid (CSF) is situated between the arachnoid and the pia.

3. The **pia mater** intimately covers the brain and the roots of cranial nerves.

C. The **dural venous sinuses** are endothelium-lined channels that receive cerebral veins; dural sinuses lack smooth muscle and valves (Figure 8–3).

1. **Cerebral veins** are known as **bridging veins** because they traverse the subdural space (a potential space) between the arachnoid and the meningeal dura to drain into dural venous sinuses.

SUBDURAL HEMATOMA

Skull trauma *may cause a shearing of bridging veins at points where they enter dural venous sinuses; venous blood may accumulate in the subdural space and result in a **subdural hematoma**.*

- *Venous blood that slowly accumulates in the subdural space forms a **crescent-shaped hematoma** not bound by sutures of the skull.*
- *Patients with a chronic subdural hematoma experience headache, impairment of cognitive skills, and gait instability.*

2. The **dural venous sinuses**, in particular the superior sagittal sinus, are sites of resorption of CSF from the subarachnoid space by way of arachnoid granulations that protrude into the sinuses.

3. **Emissary veins** are valveless veins that pass through openings in the skull and allow dural sinuses to communicate with extracranial veins.

4. The **superior sagittal sinus**, the **straight sinus**, and the **occipital sinus** carry cerebral venous blood toward the confluence of the sinuses (see Figure 8–3).

    a. The **confluence of the sinuses** is situated near the posterior midline deep to the occipital bone.

    b. The **straight sinus** is formed by the union of the inferior sagittal sinus and the **great cerebral vein of Galen**.

5. The **transverse sinuses** carry blood from the confluence of the sinuses into the sigmoid sinuses.

    a. Most of the blood from the **superior sagittal sinus** passes through the confluence and enters the right transverse sinus; most of the blood from the straight sinus enters the left transverse sinus.

    b. Each sigmoid sinus passes through a **jugular foramen**, joins with an inferior petrosal sinus, and drains into an internal jugular vein.

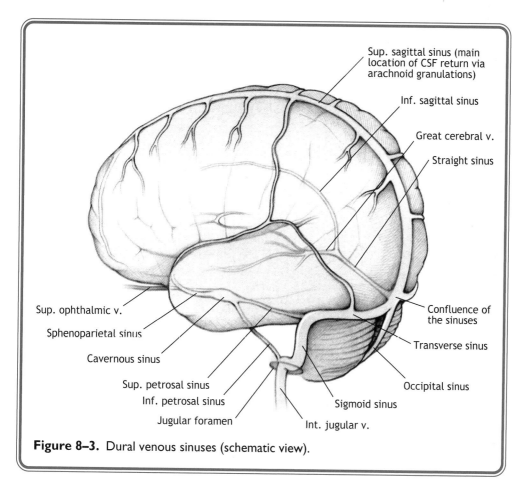

Sup. sagittal sinus (main
location of CSF return via
arachnoid granulations)

Inf. sagittal sinus

Great cerebral v.

Straight sinus

Sup. ophthalmic v.

Sphenoparietal sinus

Cavernous sinus

Sup. petrosal sinus

Inf. petrosal sinus

Jugular foramen

Confluence of
the sinuses

Transverse sinus

Occipital sinus

Sigmoid sinus

Int. jugular v.

Figure 8–3. Dural venous sinuses (schematic view).

6. The **cavernous sinuses** are dural venous sinuses situated lateral to the body of the sphenoid and the pituitary gland (Figure 8–4).
   a. The **cavernous sinuses receive venous blood** from cerebral veins and from the sphenoparietal venous sinuses.
   b. Each **cavernous sinus drains** into a superior and an inferior petrosal sinus; intercavernous sinuses interconnect the 2 cavernous sinuses.
   c. The **internal carotid artery and its periarterial plexus** of postganglionic sympathetic axons and the abducens nerve (CN VI) course through the middle of each cavernous sinus. The ophthalmic (CN V_1) and maxillary divisions (CN V_2) of the trigeminal nerve, the oculomotor nerve (CN III), and the trochlear nerve (CN IV) course in the lateral wall of each sinus.
   d. The **superior and inferior ophthalmic veins** are emissary veins that link the cavernous sinus with tributaries of the facial vein near the orbit and with the pterygoid plexus of veins in the infratemporal fossa.

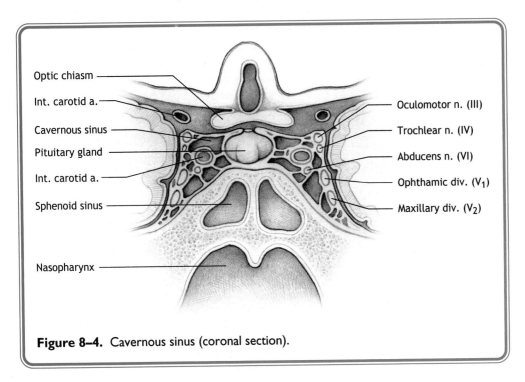

Figure 8–4. Cavernous sinus (coronal section).

Optic chiasm

Int. carotid a.

Cavernous sinus

Pituitary gland

Int. carotid a.

Sphenoid sinus

Nasopharynx

Oculomotor n. (III)

Trochlear n. (IV)

Abducens n. (VI)

Ophthamic div. (V₁)

Maxillary div. (V₂)

CLINICAL CORRELATION

THROMBOSIS OF THE CAVERNOUS SINUS

*A **thrombosis of the cavernous sinus** may result from an infection that is transported from the face to the cavernous sinus by the superior ophthalmic vein or the inferior ophthalmic vein.*

- *Patients may initially experience an **internal strabismus** resulting from a lesion of the abducens nerve.*
- *Patients may later exhibit a **loss of all ocular movements** because of oculomotor and trochlear nerve involvement and pain and numbness in the face and scalp because of involvement of the ophthalmic and maxillary divisions of the trigeminal nerve.*
- *An **infection in a cavernous sinus** may spread to the other cavernous sinus through the intercavernous sinuses.*

>   **7.** The **basilar plexus of veins** are emissary veins, which link the inferior petrosal sinuses with the internal vertebral plexus of veins in the epidural space of the vertebral canal.
>
>   **D.** The **floor of the cranial cavity** is divided into 3 shallow compartments: the anterior, middle, and posterior cranial fossae, which contain foramina and fissures that transmit blood vessels, cranial nerves, and their branches (Table 8–1 and Figure 8–5).

III. The **face and scalp** are formed by bones of the facial skeleton.

  A. The **skeletal muscles** of the face and scalp include muscles of facial expression and mastication.

    1. Muscles of facial expression are innervated by the facial nerve (CN VII) and act as sphincters and dilators of openings on the face.

Table 8–1. Major foramina of the skull and transmitted structures.

Bone	Opening	Transmits
Anterior cranial fossa		
Ethmoid bone	Cribriform plate	Olfactory nerves (I)
	Anterior ethmoidal foramen	Anterior ethmoidal nerve (nasociliary n. br.) and vessels (ophthalmic a. and v.)
	Posterior ethmoidal foramen	Posterior ethmoidal nerve (nasociliary n. br.) and vessels (ophthalmic a. and v.)
Middle cranial fossa		
Sphenoid bone	Optic canal	Optic nerve (II); ophthalmic artery
	Superior orbital fissure	Oculomotor nerve (III), trochlear nerve (IV), abducens nerve (VI), branches of ophthalmic division (V_1) of trigeminal nerve (V) (nasociliary, frontal and lacrimal n.) Superior ophthalmic vein
	Foramen rotundum	Maxillary division (V_2) of trigeminal nerve (V)
	Foramen ovale	Mandibular division (V_3) and motor root of trigeminal nerve (V) Lesser petrosal nerve [br. of glossopharyngeal nerve (IX)]
	Foramen spinosum	Middle meningeal artery (br. of maxillary a.) Nervus spinosus (meningeal branch of V3)
Temporal	Carotid canal	Internal carotid artery Internal carotid plexus (postganglionic sympathetic axons from superior cervical ganglion)
	Hiatus of facial canal	Greater petrosal nerve (VII)
	Hiatus for lesser petrosal n.	Lesser petrosal nerve (IX)
Juncture between the sphenoid, temporal, and occipital bones	Foramen lacerum	A canal of about 1 cm in length; inferior aperture closed in life by cartilage Carotid canal opens into lacerum posterolaterally; pterygoid canal opens into lacerum anteriorly Internal carotid artery and internal carotid plexus cross the width of the foramen Greater petrosal n. (from VII) and deep petrosal n. (from carotid plexus) unite in lacerum to form nerve of pterygoid canal

continued

Table 8–1. Major foramina of the skull and transmitted structures. (*continued*)

Bone	Opening	Transmits
Posterior cranial fossa		
Temporal bone	Internal auditory meatus	Vestibulocochlear nerve (VIII), Facial nerve (VII), labyrinthine artery
Juncture between temporal and occipital bones	Jugular foramen	Anterior compartment: inferior petrosal sinus Intermediate compartment: Glossopharyngeal nerve (IX), Vagus nerve (X), Spinal accessory nerve (XI) Posterior compartment: sigmoid sinus
Occipital bone	Hypoglossal canal Foramen magnum	Hypoglossal nerve (XII) Spinal cord, vertebral arteries, spinal roots of both Accessory (XI) nerves
Other openings		
Temporal bone (base of skull)	Stylomastoid foramen Petrotympanic fissure	Facial nerve (VII) Chorda tympani n. of CN VII
Frontal bone	Supraorbital foramen	Supraorbital nerve, (V_1), supraorbital a. and v.
Maxilla	Infraorbital foramen Incisive foramen Pterygomaxillary fissure	Infraorbital nerve, (V_2), infraorbital a. and v. Nasopalatine nerve and a. Posterior superior alveolar n. (V_2), maxillary a.
Junction between perpendicular process of palatine and sphenoid bone	Sphenopalatine foramen	Nasopalatine nerve (V_2), sphenopalatine a. and v.
Palatine bone	Greater palatine foramen Lesser palatine foramen	Greater palatine nerve (V_2), greater palatine a. and v. Lesser palatine nerve, V_2 lesser palatine a. and v.
Mandible	Mandibular foramen Mental foramen	Inferior alveolar nerve V_3 inferior alveolar a. and v. Mental nerve V_3, mental a and v.

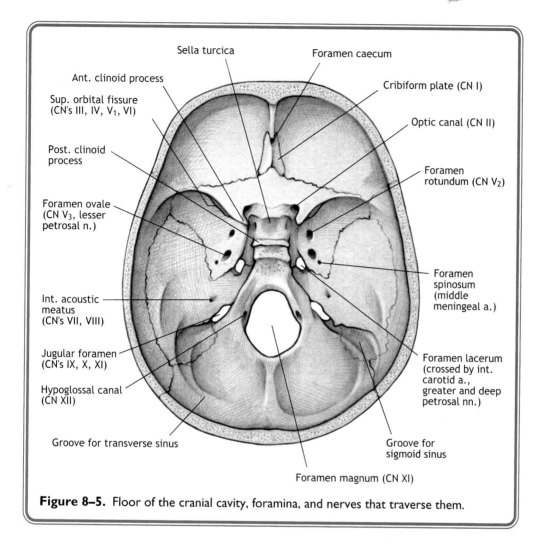

Figure 8–5. Floor of the cranial cavity, foramina, and nerves that traverse them.

2. **Muscles of mastication** are innervated by the mandibular nerve of CN V_3 and act to move the mandible at the temporomandibular joint (TMJ).

B. The **skin of the face and scalp** is innervated by cutaneous branches of the 3 divisions of the trigeminal nerve and by cervical spinal nerves (Figure 8–6).

1. **Branches of the ophthalmic division** of CN V_1 innervate the scalp anterior to the vertex, including skin of the upper eyelid, dorsum, and tip of the nose.

2. **Branches of the maxillary division** (CN V_2) innervate the cheek, including skin covering the zygomatic arch and maxilla, the lower lid, and the upper lip.

3. **Branches of the mandibular division** (CN V_3) innervate the lateral aspect of the scalp, including skin of the lateral face anterior to the external auditory meatus, skin covering the mandible, and skin of the lower lip.

4. **Branches of cervical spinal nerves** innervate skin over the angle of the mandible and the scalp posterior to the vertex.

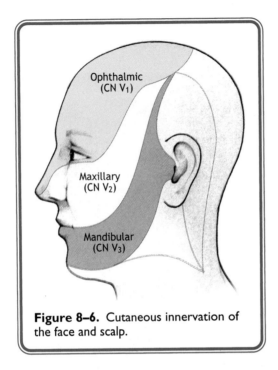

Figure 8–6. Cutaneous innervation of the face and scalp.

C. The **face and scalp are supplied** by branches of the external carotid and ophthalmic arteries.

D. The **face and scalp are drained** by tributaries of the retromandibular and facial veins.

E. The **parotid gland** is situated in the lateral part of the face on the surface of the masseter muscle. A deep part of the gland extends between the ramus of the mandible and the mastoid process.

　1. The **parotid duct** crosses the masseter, passes through the buccinator, and opens into the oral cavity near the second upper molar.

　2. The **parotid gland** is traversed by muscular branches of the facial nerve, the retromandibular vein, and the external carotid artery (Figure 8–7).

PAROTID GLAND TUMOR

*A **tumor of the parotid gland** may compress the muscular branches of the facial nerve and cause weakness of muscles of facial expression on the side of the tumor.*

　3. The parotid gland is innervated by preganglionic parasympathetic axons in the glossopharyngeal nerve and by postganglionic parasympathetic axons from the otic ganglion.

IV. The **nasal cavity** is an air-filled chamber lined anteriorly by skin and posteriorly by mucosa. The nasal cavity mucosa functions to warm, moisten, and filter inspired air (Figure 8–8).

A. The nasal cavity begins at the external nares and communicates posteriorly with the nasopharynx through the choanae.

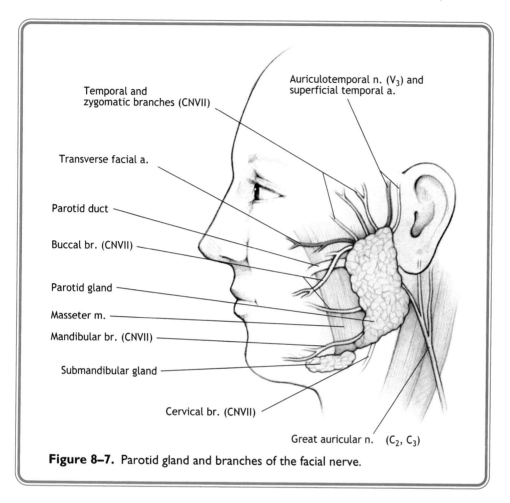

Figure 8–7. Parotid gland and branches of the facial nerve.

B. The **nasal septum** subdivides the nasal cavity in the midline, and the lateral wall of the nasal cavity contains 3 scroll-like bones, which form the superior, middle, and inferior conchae.

C. The **paranasal sinuses** (frontal, maxillary, sphenoid, and ethmoid) are areas of pneumatized bone lined by mucosa; the **mucosal secretions** of the sinuses drain mainly into meatuses, or air-filled channels, between the conchae and the lateral wall of the nasal cavity.

1. The **superior meatus** is the site of drainage of the posterior ethmoidal cells.
2. The **middle nasal meatus** is the site of drainage of the anterior and middle ethmoid air cells, the frontal sinus, and the maxillary sinus.
3. The **inferior meatus** contains the opening of the nasolacrimal duct, which drains tears from the lacrimal sac in the medial part of the orbit.
4. The **sphenoethmoidal recess** is the site of drainage of the sphenoid sinus.

D. The **maxillary sinus** is the largest paranasal sinus and is situated between the orbit and the alveolar process of the maxilla, which contains the maxillary teeth.

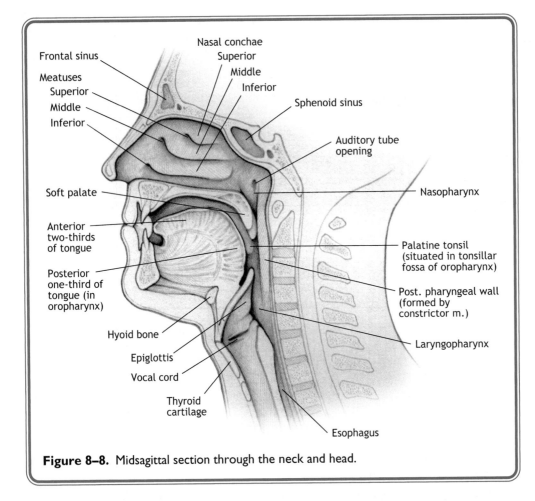

Figure 8–8. Midsagittal section through the neck and head.

SINUSITIS

*The **maxillary sinus** is a common site of **sinusitis**, an inflammation of mucosa that may result from an infection in the sinus or **obstruction of the ostium** that drains the maxillary sinus. The maxillary sinus has poor gravitational drainage because the ostium of the sinus is situated on the superior part of the medial wall of the sinus.*

    E. The **mucosa of the nasal cavity** is innervated by the olfactory nerve (CN I) and by branches of the ophthalmic (CN V_1) and maxillary (CN V_2) divisions of the trigeminal nerve.

        1. The **olfactory nerve** (CN I) is a sensory nerve that detects odors that enter the nasal cavity (Figure 8–9 and Table 8–2).

            a. The **cell bodies of bipolar olfactory neurons** are scattered in the olfactory mucosa and are not collected together in a sensory ganglion.

            b. The **olfactory nerve** consists of numerous fascicles of central processes of olfactory neurons that pass superiorly through the cribriform plate of the ethmoid and into the anterior cranial fossa.

            c. The **central processes of olfactory neurons** synapse in the olfactory bulb in the floor of the anterior cranial fossa.

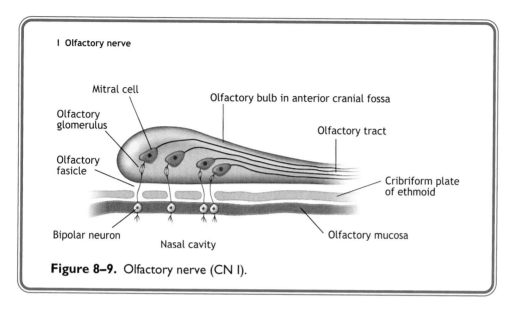

I Olfactory nerve

Mitral cell

Olfactory bulb in anterior cranial fossa

Olfactory glomerulus

Olfactory tract

Olfactory fasicle

Cribriform plate of ethmoid

Bipolar neuron

Olfactory mucosa

Nasal cavity

Figure 8–9. Olfactory nerve (CN I).

      d. **Primary olfactory neurons** are continuously replaced. The life span of these cells ranges from 30 to 120 days in mammals.

OLFACTORY NERVE LESIONS

*A lesion to the olfactory nerve fibers may alter (**hyposmia**) or distort (**dysosmia**) the sense of smell or result in a complete loss (**anosmia**).*
- *Olfactory deficits may be caused by a fracture of the cribriform plate, which damages the primary olfactory neurons.*
- *A **fracture of the cribriform plate** may also tear the meninges of the olfactory bulb and result in CSF rhinorrhea, a discharge of CSF from the nostrils.*

      2. **Branches of the ophthalmic division** (CN V_1) innervate a small part of the lateral nasal wall.

      3. **Branches of the maxillary nerve** (CN V_2) supply most of the lateral nasal wall and nasal septum.

    F. The **mucous glands of the nasal cavity** are innervated by preganglionic parasympathetics in the facial nerve and by postganglionic parasympathetic axons from the pterygopalatine ganglion.

    G. The **maxillary, ophthalmic, and facial arteries** supply the nasal cavity. Branches of all 3 vessels contribute to Kiesselbach's plexus on the anterior part of the nasal septum, which helps regulate the thermal environment of inspired air.

EPISTAXIS

*Epistaxis, or bleeding from the nose, usually results from laceration of the arteries of **Kiesselbach's plexus.***

   V. The **pterygopalatine fossa** is a space at a "crossroad" of the skull.

    A. The **fossa** communicates laterally with the infratemporal fossa, medially with the nasal cavity, anteriorly with the orbit, inferiorly with the oral cavity, and posteriorly with the middle cranial fossa and the base of the skull.

Table 8–2. Cranial nerves: Functional and clinical features.

No.	Name	Type	Function	Lesions Result In
I	Olfactory	Sensory:	Smells	Dysosmia
II	Optic	Sensory:	Sees (Optic nerve is really a tract of CNs)	Visual field deficits (anopsia) Loss of sensory limb of light reflex with III (Only nerve to be affected by multiple sclerosis)
VIII	Cochlear Vestibular	Sensory:	Hears Linear acceleration (gravity) Angular acceleration (head turning)	Sensorineural hearing loss Loss of balance Nystagmus
III	Oculomotor	Motor:	To all eyeball muscles except LR and SO; adduction (medial rectus) most important action Elevates upper eyelid (levator palpebrae superioris)	Diplopia with external strabismus Ptosis
		Parasympathetic:	Constricts pupil (sphincter pupillae) Accommodates (ciliary muscle)	Dilated pupil, loss of motor limb of light reflex with II Loss of near response
IV	Trochlear	Motor:	To superior oblique—depresses and abducts eyeball Intorts eyeball	Weakness looking down with eye adducted Difficulty reading, going down stairs Head tilt away from lesioned side
VI	Abducens	Motor:	To lateral rectus—abducts eyeball	Diplopia with internal strabismus Pseudoptosis
XI	Accessory	Motor:	Turns chin to opposite side (sternocleidomastoid) Elevates and upwardly rotates scapula (trapezius)	Weakness turning chin to opposite side, shoulder droop, difficulty combing hair
XII	Hypoglossal	Motor:	To all muscles that act on tongue except palatoglossus (X) (hyoglossus, styloglossus, genioglossus, intrinsics)	Tongue deviation on protrusion toward side of lesioned nerve

CNS, central nervous system; LR, lateral rectus; SO, superior oblique

continued

Table 8–2. Cranial nerves: Functional and clinical features. (*continued*)

No.	Name	Type	Function	Lesions Result In
V	Trigeminal Oophthalmic (V1)	Sensory:	General sensation (touch, pain, temperature) of forehead, scalp, cornea, dorsum of nose, nasal cavity	Loss of general sensation in skin of forehead/scalp Loss of sensory limb of blink reflex with VII
	Maxillary (V2)	Sensory:	General sensation of palate, nasal cavity, maxillary face, maxillary teeth	Loss of general sensation in skin over maxilla, maxillary teeth
	Mandibular (V3)	Sensory:	General sensation of anterior 2/3 tongue, mandibular face, mandibular teeth	V3—loss of general sensation in skin over mandible, mandibular teeth, tongue, weakness in chewing
		Motor:	To muscles of mastication (temporalis, masseter, medial and lateral pterygoids) Anterior belly of digastric, mylohyoid, tensor tympani, tensor palati)	Jaw deviation on protrusion toward lesioned nerve
VII	Facial	Motor:	To muscles of facial expression (orbicularis oculi and oris, platysma, buccinator) and to posterior belly of digastric, stylohyoid, stapedius	Corner of mouth droop, can't close eye, wrinkle forehead Loss of motor limb of blink reflex with V Hyperacusis
		Sensory:	Taste anterior 2/3 tongue/ soft palate General sensation in skin behind ear	Altered taste from anterior 2/3 of tongue Pain behind the ear Reduction in output of saliva
		Parasym-pathetic:	Saliva (submandibular, sublingual glands) Tears (lacrimal gland) Mucus (nasal and palatine glands)	Reduction in secretions Eye dry and red.
IX	Glosso-pharyngeal	Motor: Sensory:	To one muscle-stylopharyngeus General sensation of oropharynx, carotid sinus and body Taste and general sensation posterior 1/3 of tongue	Loss of sensory limb of gag reflex with X
		Parasym-pathetic:	Saliva (parotid gland)	Reduction in output of saliva

continued

Table 8–2. Cranial nerves: Functional and clinical features. (*continued*)

No.	Name	Type	Function	Lesions Result In
X	Vagus	Motor:	To muscles of palate except tensor palati (V) droop	Nasal speech, nasal regurgitation, palate
			To muscles of pharynx except stylopharyngeus (IX)	Tip of uvula deviates away from lesioned nerve
			To all muscles of larynx	Dysphagia, loss of motor limb of gag reflex with IX
				Hoarseness
		Sensory:	Senses larynx and laryngopharynx	Loss of cough reflex
		Parasympathetic:	To cardiac muscle and lung smooth muscle	
			To glands and smooth muscle in foregut and midgut	Reduced peristalsis
	Sympathetics to face, scalp, orbit	Motor	Elevates upper eyelid (superior tarsal muscle)	Ptosis
			Dilates pupil (dilator pupillae)	Constricted pupil (miosis),
			Innervates sweat glands of face and scalp	Loss of sweating (anhydrosis) (Horner's syndrome)

B. The **pterygopalatine fossa** contains the pterygopalatine ganglion and preganglionic and postganglionic parasympathetic axons associated with it.
1. The **preganglionic parasympathetic axons** course in the greater petrosal nerve, a branch of CN VII (see later discussion).
2. The **postganglionic parasympathetic axons** supply mucous glands of the nasal cavity, oral cavity, nasopharynx, and the lacrimal gland.
C. Two **fissures** (pterygomaxillary and inferior orbital), 2 **foramina** (rotundum and sphenopalatine), and 2 **major canals** (the pterygoid and palatine) open into the **pterygopalatine fossa**. These openings transmit branches of the maxillary artery, maxillary nerve, and facial nerve into or out of the fossa.
1. **Branches of the maxillary artery** that traverse the fossa supply the nasal cavity, hard and soft palate, nasopharynx, and maxillary teeth and face.
2. The **maxillary division of the trigeminal nerve** (V$_2$) contains fibers that convey general sensations of touch, pain, and temperature from skin, mucosa, and dura. The neuronal cell bodies of the touch, pain, and temperature neurons of all 3 divisions of the trigeminal nerve are found in the trigeminal ganglion in the middle cranial fossa (Figure 8–10).
3. **All 3 divisions of the trigeminal nerve** also carry proprioceptive fibers. The cell bodies of the proprioceptive neurons are found inside the central nervous system (CNS); these neurons innervate sensory receptors in skeletal muscles innervated by cranial nerves.
   a. The **maxillary division** traverses the foramen rotundum, becomes the maxillary nerve in the pterygopalatine fossa, and gives rise to 3 main branches, the zygomatic, infraorbital, and pterygopalatine nerves (a mnemonic for these nerves is "ZIP").

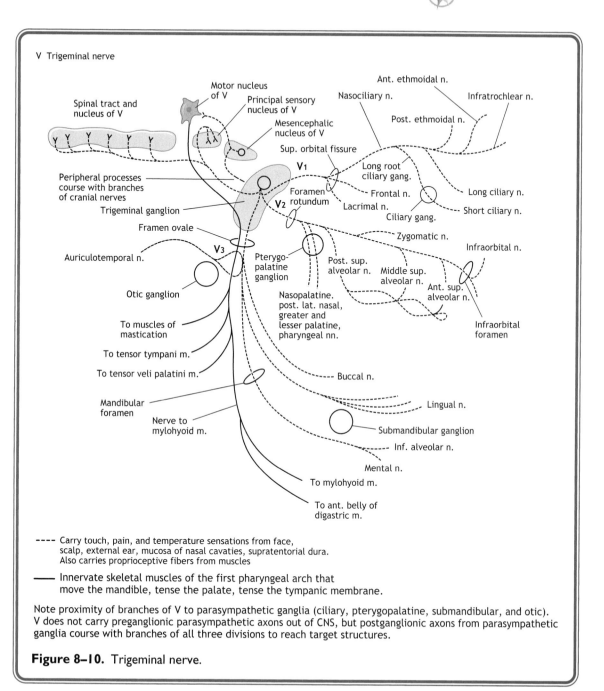

V Trigeminal nerve

Figure 8–10. Trigeminal nerve.

---- Carry touch, pain, and temperature sensations from face,
scalp, external ear, mucosa of nasal cavaties, supratentorial dura.
Also carries proprioceptive fibers from muscles

—— Innervate skeletal muscles of the first pharyngeal arch that
move the mandible, tense the palate, tense the tympanic membrane.

Note proximity of branches of V to parasympathetic ganglia (ciliary, pterygopalatine, submandibular, and otic).
V does not carry preganglionic parasympathetic axons out of CNS, but postganglionic axons from parasympathetic
ganglia course with branches of all three divisions to reach target structures.

b. The **pterygopalatine nerves** pass through the pterygopalatine ganglion (without synapsing) and become renamed as they emerge from the ganglion as the greater and lesser palatine nerves, nasopalatine nerve, posterior lateral nasal nerve, and pharyngeal nerve.

  4. The **pterygoid canal** contains the nerve of the pterygoid canal. This nerve is formed by the deep petrosal nerve and the greater petrosal nerve, which join in the foramen lacerum (Figure 8–11).

    a. The **greater petrosal nerve** is a mixed branch of the facial nerve that carries preganglionic parasympathetic axons that synapse in the pterygopalatine ganglion and taste fibers from the palate.

    b. The **deep petrosal nerve** is a branch of the periarterial plexus on the internal carotid artery that carries postganglionic sympathetic axons from the superior cervical ganglion.

VI. The **orbit** contains the eyeball, skeletal and smooth muscles that act on the eyeball, pupil and eyelid, superior ophthalmic vein and ophthalmic artery, and branches of 5 cranial nerves (II, III, IV, V_1, and VI).

  A. The **optic canal and the superior orbital fissure** transmit structures into and out of the orbit (see Table 8–1).

    1. The optic canal transmits the optic nerve and the ophthalmic artery.

    2. The superior orbital fissure transmits CNs III, IV, V_1, and VI and the superior ophthalmic vein (Figure 8–12).

  B. The **eyeball** mainly consists of 3 concentric layers: the **sclera, choroid,** and **retina**; anteriorly, the eyeball contains the **cornea** and the **lens**, which are the transparent refractive media of the eye.

    1. The lens lies posterior to the cornea and is separated from it by the iris and the pupil.

    2. The lens is held in place by the fibers of the suspensory ligament. Changes in the tension of the suspensory ligament alter the **refractive power** of the lens by allowing its anterior and posterior surfaces to increase or decrease their curvature. **Relaxation of the suspensory ligament** results in increased curvature of the lens.

  C. The orbit contains **6 skeletal muscles** (4 rectus, 2 oblique) that act on the eyeball and a **seventh skeletal muscle**, the levator palpebrae superioris, that elevates the upper eyelid (Figure 8–13 and Table 8–3).

    1. The **4 rectus muscles** have an attachment to a common tendinous ring at the back of the orbit; the superior oblique attaches to the sphenoid bone above the tendinous ring, and the inferior oblique attaches to the orbital plate of the maxilla in the anterior part of the orbit.

    2. All muscles that move the eyeball except for the inferior oblique traverse the length of the orbit.

    3. The **medial and lateral rectus muscles** adduct and abduct the eyeball, respectively. The superior and inferior rectus muscles elevate and depress the eyeball, respectively, and adduct the eyeball. The superior and inferior oblique muscles depress and elevate the eyeball, respectively, and abduct the eyeball. The 2 superior muscles also medially rotate the eyeball or twist it inwardly (intorted); the 2 inferior muscles extort the eyeball (Table 8–6 and Figure 8–13).

    4. To test individual ocular muscles, the eyeball is moved to a position that isolates the muscle being tested (see Figure 8–13).

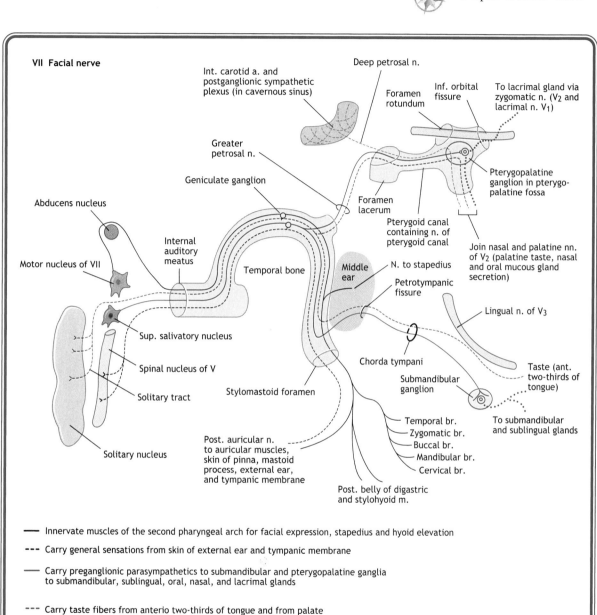

VII Facial nerve

— Innervate muscles of the second pharyngeal arch for facial expression, stapedius and hyoid elevation

--- Carry general sensations from skin of external ear and tympanic membrane

— Carry preganglionic parasympathetics to submandibular and pterygopalatine ganglia to submandibular, sublingual, oral, nasal, and lacrimal glands

--- Carry taste fibers from anterio two-thirds of tongue and from palate

Figure 8–11. Facial nerve.

    a. The medial and lateral rectus muscles are tested by asking the patient to adduct or abduct the eyeball, respectively.

    b. The superior and inferior rectus muscles are tested for elevation and depression, respectively, when the eyeball is abducted.

    c. The superior and inferior oblique muscles are tested for depression and elevation, respectively, when the eyeball is adducted.

5. Four of the 6 muscles—the superior, inferior and medial rectus, and inferior oblique—are innervated by the oculomotor nerve (CN III) (see Figure 8–12).

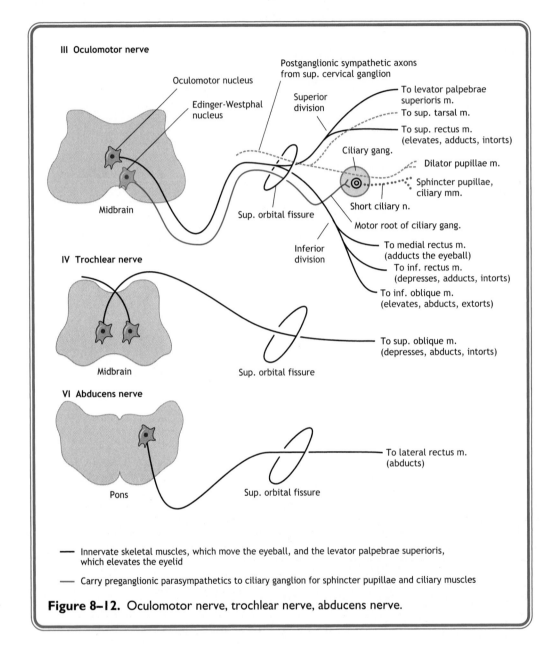

Figure 8–12. Oculomotor nerve, trochlear nerve, abducens nerve.

6. The **superior oblique muscle** is innervated by the trochlear nerve (CN IV), and the lateral rectus is innervated by the abducens nerve (CN VI) (see Figure 8–12).

7. The **skeletal muscle part of the levator palpebrae superioris** is innervated by the oculomotor nerve (CN III).

D. The orbit contains **4 smooth muscles**.

1. The **superior tarsal muscle** is the smooth muscle part of the levator palpebrae superioris, which elevates the upper eyelid.

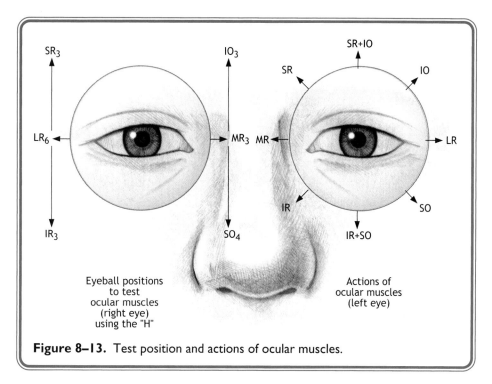

Figure 8–13. Test position and actions of ocular muscles.

Table 8–3. Muscles of the orbit.

Action	Muscles Involved	Innervation
Elevates and adducts eyeball (cornea moves up and in)	Superior rectus	Oculomotor nerve
Depresses and adducts eyeball (cornea moves down and in)	Inferior rectus	Oculomotor nerve
Adducts eyeball	Medial rectus	Oculomotor nerve
Abducts eyeball	Lateral rectus	Abducens nerve
Elevates and abducts eyeball (cornea moves up and out)	Inferior oblique	Oculomotor nerve
Depresses and abducts eyeball (cornea moves down and out)	Superior oblique	Trochlear nerve
Elevates eyelid	Levator palpebrae superioris	Oculomotor nerve
Smooth Muscles		
Elevates eyelid in sympathetic response	Superior tarsal	Postganglionic sympathetic axons from superior cervical ganglion
Dilates pupil	Dilator pupillae	Postganglionic sympathetic axons from superior cervical ganglion

continued

Table 8–3. Muscles of the orbit. (*continued*)

Action	Muscles Involved	Innervation
Constricts pupil	Constrictor pupillae	Oculomotor nerve
Relaxes suspensory ligament to permit lens to round up in near response	Ciliary	Oculomotor nerve

    2. The **dilator pupillae and the constrictor pupillae** are in the anterior aspect of the eyeball in the iris and act to alter the diameter of the pupil.

    3. The **ciliary muscle** is in the ciliary body; when the ciliary muscle contracts, the suspensory ligaments relax, allowing the lens to "round up" for near vision

  E. In the **accommodation reflex**, or the near response, the ciliary muscle contracts and the lens becomes thicker, and there is constriction of the pupils and convergence. **Convergence** results from simultaneous contraction of both medial rectus muscles. All of the muscles used in the near response are innervated by parasympathetic or skeletal motor fibers in the oculomotor (CN III) nerve.

  F. The **pupillary light reflex** uses sensory fibers of the optic nerve (CN II) and parasympathetic fibers of the oculomotor nerve (CN III). Under normal conditions, presentation of light to one eye should result in a constriction of both pupils.

PUPILLARY REFLEX DEFICITS

CLINICAL CORRELATION

Deficits in the response of 1 or both pupils to light may be caused by a lesion of either the afferent or efferent components of the light reflex.

An **afferent pupillary defect** (a Marcus Gunn pupil) may result from **lesions to the optic nerve** and can be confirmed in patients by the swinging flashlight test. When the normal eye is exposed to light, both pupils constrict; however, when the flashlight is swung to the affected eye, both pupils paradoxically dilate. **Lesions to the oculomotor nerve** may result in an efferent pupillary defect. In these patients, the pupil is dilated on the affected side and does not constrict in response to light shone in either eye. If the lesion is complete, the patient may have a "blown pupil."

  G. The **lacrimal gland** is located in the upper lateral corner of the orbit. It is innervated by preganglionic parasympathetics in CN VII and postganglionic parasympathetic axons from the pterygopalatine ganglion.

  H. The **optic nerve** (CN II) is not actually a nerve but a tract of the brain containing sensory axons.

    1. **Retinal ganglion cells** give rise to axons that form the optic nerve (Figure 8–14).

    2. All of the **myelin of axons of the optic nerve** are formed by **oligodendrocytes;** consequently, the optic nerve is susceptible to **CNS demyelinating diseases** such as **multiple sclerosis**, which typically do not affect other cranial or spinal nerves.

    3. The myelinated axons of the optic nerve are ensheathed by the **3 layers of meninges** (meningeal dura, arachnoid, and pia) that surround the rest of the CNS. The **subarachnoid space** containing CSF extends along the optic nerve to the optic disk at the back of the eyeball.

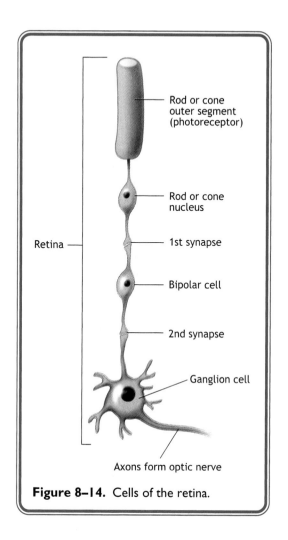

Rod or cone
outer segment
(photoreceptor)

Rod or cone
nucleus

Retina

1st synapse

Bipolar cell

2nd synapse

Ganglion cell

Axons form optic nerve

Figure 8–14. Cells of the retina.

PAPILLEDEMA

*An **increase in intracranial CSF pressure** may result in **papilledema**, a swelling of the nerve at the optic disk caused by reduced venous return from the retina.*

    4. At the **optic disk**, each optic nerve leaves the eyeball and courses posteromedially into the middle cranial fossa through the optic canal.

        a. The optic nerve fibers converge in the midline anterior and superior to the pituitary gland to form the optic chiasm. They then diverge to form the optic tracts (Figure 8–15).

        b. At the chiasm, approximately 60% of the optic nerve fibers cross the midline; 40% remain uncrossed.

        c. The optic nerve fibers that cross in the chiasm arise from ganglion cells in the nasal half of each retina but convey visual information entering the eye from the temporal half of each visual field.

        d. Optic nerve fibers that do not cross in the chiasm arise from ganglion cells in the temporal half of each retina, which convey visual information from each nasal half of each visual field (see Figure 8–15).

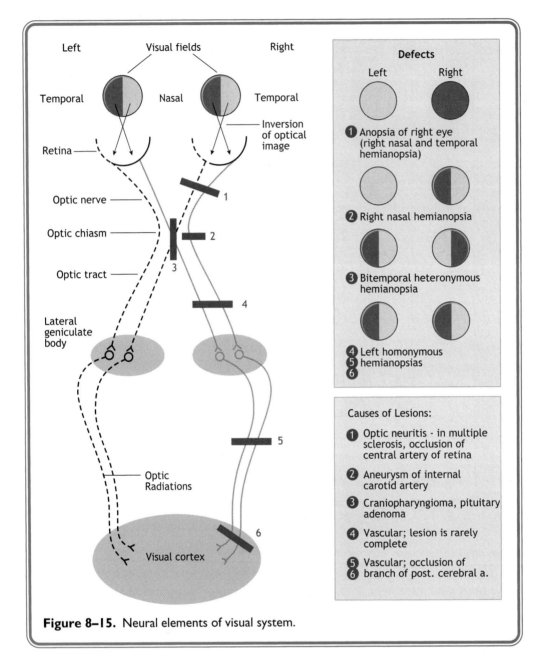

Left Visual fields Right

Temporal Nasal Temporal

Inversion of optical image

Retina

Optic nerve — 1

Optic chiasm — 2

3

Optic tract

4

Lateral geniculate body

Optic Radiations

5

Visual cortex 6

Defects

Left Right

❶ Anopsia of right eye (right nasal and temporal hemianopsia)

❷ Right nasal hemianopsia

❸ Bitemporal heteronymous hemianopsia

❹ Left homonymous
❺ hemianopsias
❻

Causes of Lesions:

❶ Optic neuritis - in multiple sclerosis, occlusion of central artery of retina

❷ Aneurysm of internal carotid artery

❸ Craniopharyngioma, pituitary adenoma

❹ Vascular; lesion is rarely complete

❺ Vascular; occlusion of
❻ branch of post. cerebral a.

Figure 8–15. Neural elements of visual system.

5. The **optic tracts** contain a combination of crossed and uncrossed optic nerve fibers.
   a. There are no synapses in the optic nerves, chiasm, or tracts; therefore, axons of the ganglion cells of the retina extend uninterrupted to synapse mainly in the lateral geniculate nucleus of the thalamus for primary visual processing.
   b. Axons of cells in the lateral geniculate nucleus form the visual radiations that project to primary visual cortex.

VISUAL FIELD DEFECTS

Lesions of the optic nerve, chiasm, tract, and visual radiations *result in visual field deficits (see Figure 8–15).*

*   *Complete nasal and temporal visual field deficits are **anopsias**. Deficits in a nasal or temporal half of a visual field are hemianopsias. Deficits in a quadrant of a temporal or nasal visual field are **quadrantanopsias**.*
*   *The visual system inverts and reverses optical images like a camera. Therefore, lesions in parts of the visual pathway that convey information originating in the nasal half of the retina will result in deficits in a temporal visual hemifield. Lesions in parts of the visual pathway that convey information originating in the temporal half of the retina will result in deficits in a nasal visual hemifield.*
*   *Lesions in the retina produce **scotomas**, small spot-like deficits in a part of a temporal or nasal visual field of that eye.*
*   *A complete lesion of an optic nerve results in a **monocular anopsia**; eg, a lesion of the right optic nerve results in **anopsia of the right eye**.*
*   *Complete lesions of the optic chiasm **produce bitemporal heteronymous hemianopsias**. The defects are heteronymous because the deficits are in the left temporal and in the right temporal hemifields.*
*   *Lesions in visual pathways past the chiasm produce **contralateral and binocular homonymous deficits**, in the same part of a left or right nasal and temporal hemifield. Therefore, a complete lesion of the right optic tract will result in a left homonymous hemianopsia (see Figure 8–15).*

I. The **oculomotor nerve** (CN III) contains preganglionic parasympathetic and skeletal motor axons (see Figure 8–12).

  1. The oculomotor nerve (CN III) exits ventrally from the brainstem, courses in the lateral wall of the cavernous sinus, and splits into superior and inferior divisions that enter the orbit through the superior orbital fissure.

  2. The **divisions of the oculomotor nerve** innervate all of the skeletal muscles that act on the eyeball except the lateral rectus and superior oblique; the superior division also innervates the levator palpebrae superioris.

  3. The **preganglionic parasympathetic axons** course in the inferior division of CN III.

  a. The preganglionic parasympathetic axons leave the inferior division in a motor root and synapse in the ciliary ganglion.

  b. Postganglionic parasympathetic axons enter the posterior aspect of the eyeball in short ciliary nerves that innervate the sphincter pupillae and ciliary muscles.

  4. **The fibers in the oculomotor nerve** are arranged such that the parasympathetic fibers course dorsolaterally in the peripheral part of the nerve.

OCULOMOTOR NERVE LESIONS

*External **lesions of the oculomotor nerve** may result from compression by a herniated part of a hemisphere or by a **berry aneurysm**. These lesions tend to affect the parasympathetic fibers first, resulting in a dilated pupil (**mydriasis**) and a suppression of the pupillary light reflex.*

*A complete **lesion of the skeletal motor fibers of the oculomotor nerve** results most dramatically in an inability to adduct the eyeball (see Table 8–2).*

*   *Patients with an oculomotor nerve lesion may have an **external strabismus**, a laterally deviated eyeball that results from unopposed contractions of the lateral rectus and superior oblique.*
*   *Oculomotor nerve lesions may also result in a **ptosis**, which results from a weakness of the skeletal motor part of the levator palpebrae superioris muscle.*

J. The **trochlear nerve** (CN IV) contains skeletal motor axons (see Figure 8–12).
1. The trochlear nerve exits from the dorsal aspect of the brainstem, courses through the lateral wall of the cavernous sinus, and enters the orbit through the superior orbital fissure.
2. It is the only completely crossed cranial nerve and the only cranial nerve to arise from the dorsal aspect of the brain.
3. The trochlear nerve innervates a single skeletal muscle, the superior oblique.

TROCHLEAR NERVE LESIONS

A **lesion of the trochlear nerve** results in **diplopia** when a patient attempts to depress the adducted eye (see Table 8–2).
- Patients may experience difficulty in reading or difficulty in going down stairs.
- Patients with a trochlear nerve lesion may tilt their head away from the side of the lesioned nerve; this results from a weakness in the ability to rotate the affected eyeball inward (intorsion). The head tilt is an attempt to counteract the extorsion by the unopposed inferior oblique and inferior rectus muscles.
- A head tilt observed in patients with a trochlear nerve lesion might be mistaken for **torticollis**, which is caused by abnormal contractions of the sternocleidomastoid muscle.

K. The **abducens nerve** (CN VI) contains skeletal motor fibers (see Figure 8–13).
1. The abducens nerve exits from the brain at the junction of the pons and medulla, pierces the dura at the posterior aspect of the cavernous sinus, and courses through the sinus between its lateral wall and the internal carotid artery.
2. It enters the orbit through the superior orbital fissure and innervates the lateral rectus muscle.

ABDUCENS NERVE LESIONS

Lesions of the abducens nerve result in a weakness in the ability to fully abduct the eye. The superior and inferior oblique muscles act to partially abduct the eye. Patients with an abducens nerve lesion may have an **internal strabismus**, an eyeball that is deviated medially because of the unopposed contractions of the medial rectus muscle and other adductors innervated by CN III.

The **abducens nerve** is most commonly the first nerve to be affected in a **thrombosis of the cavernous sinus** (see Figure 8–4).

L. The orbit is traversed by branches of the ophthalmic division of CN V; these fibers convey general sensations of touch, pain, and temperature from skin of the face and scalp and mucosa of the nasal cavity (see Figure 8–10).
1. The ophthalmic division gives off a meningeal branch, which innervates the superior aspect of the dura of the tentorium cerebelli and dura of the anterior cranial fossa.
2. The **ophthalmic division** courses in the lateral wall of the cavernous sinus and divides into the **nasociliary, frontal, and lacrimal nerves** (a mnemonic for these nerves is "NFL") just posterior to the superior orbital fissure.
   a. The **nasociliary nerve** branches into long ciliary, infratrochlear, anterior ethmoidal, and posterior ethmoidal nerves. Long ciliary nerves enter the posterior aspect of the eyeball and innervate the cornea. Corneal branches may also pass through the ciliary ganglion and course in short ciliary nerves.
   b. The **frontal nerve** traverses the superior orbital fissure and divides into supraorbital and supratrochlear nerves near the anterior margin of the orbit; these nerves innervate skin of the upper eyelid and scalp.

c. The **lacrimal nerve** traverses the superior orbital fissure and the length of the orbit to supply skin lateral to the orbit. The lacrimal nerve also conveys hitchhiking postganglionic parasympathetic axons from the pterygopalatine ganglion to the lacrimal gland.

VII. The **oral** cavity begins anteriorly at the vestibule and is continuous posteriorly with the oropharynx (see Figure 8–8).

A. The oral cavity is bounded anteriorly and laterally by the teeth, superiorly by the hard and soft palate, and inferiorly by the mylohyoid muscles and the tongue.

B. The **vestibule** is a horseshoe-shaped space between the teeth and the mucosal lining of the cheeks and lips.
1. **Contractions of the orbicularis oris muscle** close the anterior opening of the vestibule.
2. The **cheeks** form the lateral walls of the vestibule and contain the **buccinator muscles,** which act to compress the cheek against the molar teeth to keep food between the teeth during chewing.
3. The **parotid duct** opens into the vestibule opposite the second upper molar tooth after passing through the buccinator.

C. The **palatoglossal arches** form the posterolateral boundary between the oral cavity and the oropharynx and contain the palatoglossus muscles.

D. The **floor of the oral cavity** contains the anterior two-thirds of the tongue.
1. The **tongue** is divided by the sulcus terminalis into an anterior two-thirds, which is in the oral cavity, and a posterior one-third, which is in the oropharynx.
2. The **mucosa of the tongue** contains receptors for general sensations (touch, pain, temperature) and taste.
   a. The mucosa of the anterior two-thirds of the tongue has a dual sensory innervation. General sensations are carried by the lingual nerve of CN V_3, and taste sensations are carried by the chorda tympani of the facial nerve (CN VII).
   b. Both taste and general sensations from the posterior one-third of the tongue are carried by the lingual branch of the glossopharyngeal nerve (CN IX).
   c. The mucosa at the root of the tongue (in front of the epiglottis) is innervated by the internal laryngeal branch of the vagus nerve (CN X).
3. The **inferior surface of the tongue** contains the openings of the ducts of the submandibular and sublingual glands. These glands are innervated by preganglionic parasympathetic axons of CN VII and by postganglionic parasympathetic axons from the submandibular ganglion.
4. The tongue consists of **intrinsic muscles**, which act to alter its shape, and **4 pairs of extrinsic muscles**—genioglossus, hyoglossus, styloglossus, and palatoglossus—that act to move the tongue (Table 8–4).
5. All of the **muscles of the tongue** except the palatoglossus are innervated by the hypoglossal nerve (CN XII). The palatoglossus muscle is innervated by the vagus nerve (CN X).

E. The **hypoglossal nerve** (CN XII) contains skeletal motor axons (Figure 8–16).
1. The hypoglossal nerve exits the cranial cavity through the hypoglossal canal and crosses the proximal parts of the internal and external carotid arteries in the neck.

Table 8–4. Muscles of the tongue.

Action	Muscles Involved	Innervation
Protrudes and retracts tongue	Genioglossus	Hypoglossal nerve
Depresses tongue	Hyoglossus	Hypoglossal nerve
Elevates tongue	Styloglossus	Hypoglossal nerve
Elevates tongue	Palatoglossus	Vagus nerve (pharyngeal plexus)

2. The hypoglossal nerve enters the floor of the mouth between the hyoglossus and mylohyoid muscles.

HYPOGLOSSAL NERVE LESIONS

Lesions of the hypoglossal nerve result in a deviation of the tongue toward the side of the injured nerve on protrusion. Patients with a hypoglossal nerve lesion may experience dysarthria and difficulty moving a bolus of food from the oral cavity into the oropharynx.

    F. The **mucosa of the hard and soft palate** is innervated by palatine branches of the maxillary nerve (CN V$_2$). Taste sensations from the soft palate enter the CNS in the facial nerve (CN VII).

VIII. The **infratemporal fossa** is medial to the ramus of the mandible and the zygomatic arch. It contains the TMJ, muscles of mastication, branches of the maxillary artery, branches of the mandibular nerve (CN V$_3$), and the chorda tympani (CN VII).

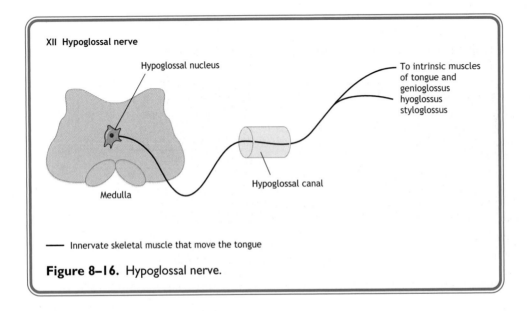

Figure 8–16. Hypoglossal nerve.

A. At the TMJ, the head (condylar process) of the mandible articulates with the mandibular fossa of the temporal bone (see Figures 8–1, 8–17a).

1. The TMJ contains an articular disk, which subdivides the joint into 2 joint cavities.

2. The superior part of the joint cavity between the disk and the mandibular fossa is a gliding joint; the mandible can be protruded and retruded at the gliding joint.

3. The inferior part of the joint cavity between the disk and the head of the mandible is a hinge joint; the mandible can be elevated and depressed at the hinge joint.

4. The combined actions at both joint cavities permit movement of the mandible about a coronal axis through the mandibular foramina.

B. There are **4 muscles of mastication**. Three of the 4—the temporalis, lateral pterygoid, and medial pterygoid—are in the infratemporal fossa. The masseter is lateral to the infratemporal fossa (Figure 8–17b).

1. The major actions of muscles of mastication are to **elevate, protrude,** and **retrude** the mandible (Table 8–5).

2. All of the **muscles of mastication** are innervated by branches of the mandibular nerve of CN V_3.

C. The **maxillary artery** courses through the infratemporal fossa and then enters the pterygopalatine fossa through the pterygomaxillary fissure (Figure 8–18).

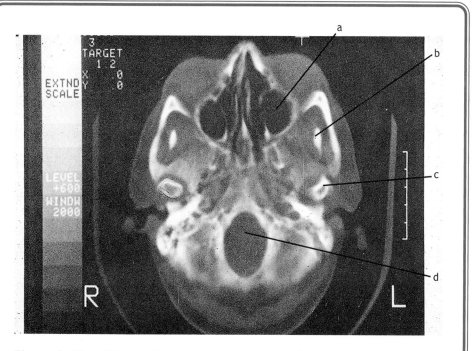

Figure 8–17a. CT scan of cross section through skull at level of temporomandibular joint. a. maxillary sinus, b. coronoid process, c. head of mandible, d. for magnum.

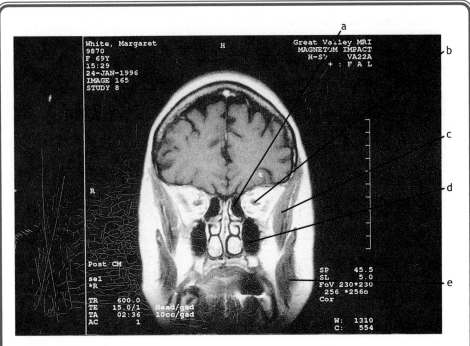

Figure 8–17b. MRI of coronal section through skull. a. ethmoid air cells, b. optic nerve, c. temporalis m., d. maxillary sinus, e. masseter m.

Table 8–5. Muscles that act at the temporomandibular joint.

Action	Muscles Involved	Innervation
Depression of mandible (open mouth)	Digastric (anterior belly) Infrahyoid (strap muscles)	Nerve to Mylohyoid (V_3) Ansa cervicalis or C1 fibers
Elevation (close mouth)	Temporalis Masseter Medial pterygoid	Mandibular nerve (V_3) Mandibular nerve (V_3) Mandibular nerve (V_3)
Protrusion	Lateral pterygoid Masseter Medial pyterygoid	Mandibular nerve (V_3) Mandibular nerve (V_3) Mandibular nerve (V_3)
Retrusion	Temporalis (posterior fibers)	Mandibular nerve (V_3)
Grinding (side-to-side movements)	Temporalis of same side Pterygoids of opposite side	Mandibular nerve (V_3) Mandibular nerve (V_3)

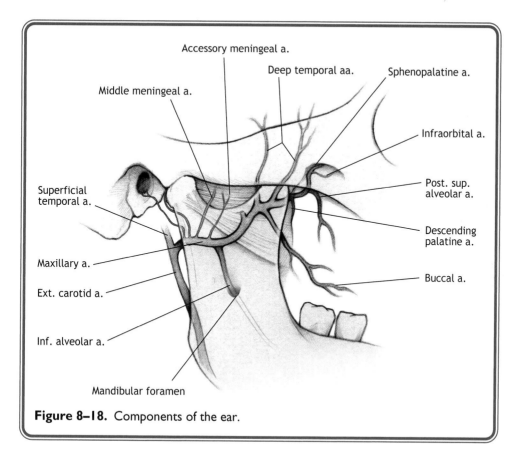

Figure 8–18. Components of the ear.

1. The **main branches** of the maxillary artery in the infratemporal fossa are the **middle meningeal artery,** the **inferior alveolar artery,** and the **deep temporal arteries.**
2. The middle meningeal artery passes through the foramen spinosum and enters the middle cranial fossa to supply the skull and dura; a **lateral skull fracture** may lacerate the middle meningeal artery and cause an epidural hematoma.

D. The **roof of the infratemporal fossa** contains the foramen ovale, foramen spinosum, and petrotympanic fissure.
   1. The **foramen ovale** transmits the mandibular division and motor root of the trigeminal nerve (V_3) and the lesser petrosal nerve (of CN IX). The lesser petrosal nerve contains preganglionic parasympathetic axons that synapse in the otic ganglion; the otic ganglion is in the infratemporal fossa just below the foramen ovale, and its axons innervate the parotid gland.
   2. The **foramen spinosum** transmits the middle meningeal artery and the meningeal branch of CN V_3.
   3. The **petrotympanic fissure** communicates with the middle ear cavity and transmits the chorda tympani of CN VII. The **chorda tympani** contains taste fibers from the anterior two-thirds of the tongue and preganglionic

parasympathetic axons that synapse in the submandibular ganglion; the submandibular ganglion is in the floor of the mouth.

E. The **mandibular nerve** (CN V₃) contains fibers that convey general sensations of touch, pain, and temperature. It also contains skeletal motor axons (see Figure 8–10).

1. The mandibular nerve is formed just below the foramen ovale by the union of the mandibular division with the motor root.

2. Muscular branches of the mandibular nerve supply the 4 muscles of mastication (masseter, temporalis, lateral and medial pterygoids) and the tensor tympani and tensor veli palatini (the only 2 tensors in the head). The nerve to the mylohyoid, the longest muscular branch, supplies the mylohyoid and anterior belly of the digastric.

3. The **main general sensory branches** of the mandibular nerve include a **meningeal branch**, and 4 cutaneous or mucosal branches—the **buccal nerve**, the **auriculotemporal nerve**, the **inferior alveolar nerve**, and the **lingual nerve** (a mnemonic word for these nerves is "BAIL"; Figure 8–10).

a. The lingual nerve also carries hitchhiking taste and parasympathetic fibers of the chorda tympani of CN VII.

b. The auriculotemporal nerve also carries hitchhiking postganglionic parasympathetic axons from cell bodies in the otic ganglion to the parotid gland.

MANDIBULAR NERVE LESIONS

CLINICAL CORRELATION

Trigeminal neuralgia (tic douloureux) *is characterized by episodes of sharp, stabbing pain that radiates over the areas innervated by sensory branches of the maxillary or mandibular divisions of the trigeminal nerve.*

- *In many patients with trigeminal neuralgia, pain radiates over the mandible extending around the TMJ and deep to the external ear.*
- *In other patients, pain radiates up the nostril and around the inferior aspect of the orbit.*
- *The pain associated with the neuralgia is frequently triggered by moving the mandible, smiling or yawning, or by cutaneous or mucosal stimulation. Trigeminal neuralgia may be caused by pressure on or interruption of the blood supply to the trigeminal ganglion.*
- *A **lesion of the motor root of the mandibular nerve** may result in a weakness of muscles of mastication and a deviation of the jaw on protrusion toward the side of the injured nerve.*

IX. The **palate** separates the oral cavity from the nasal cavity and the nasopharynx from the oropharynx (see Figure 8–8).

A. The **hard palate** is anterior. It contains the incisive foramina and the greater and lesser palatine foramina, which transmit branches of the maxillary division of CN V and branches of the maxillary artery that supply the oral cavity (see Table 8–1).

1. The **incisive foramen** transmits the nasopalatine nerve and the nasopalatine artery.

2. The **greater palatine foramen** transmits the greater palatine nerve and artery, and the lesser palatine foramen transmits the lesser palatine nerve and artery.

B. The **soft palate** is posterior. It contains skeletal muscles and ends posteriorly at the uvula.

1. The **core of the soft palate** is formed by a membranous palatine aponeurosis.
2. Four pairs of skeletal muscles—the palatopharyngeus, palatoglossus, levator veli palatini, and tensor veli palatini—attach to the **palatine aponeurosis** and act on the soft palate (see Table 8–6).
3. All the **muscles of the soft palate** except for the tensor veli palatini are innervated by pharyngeal branches of the vagus nerve (CN X). The tensor veli palatini muscle is innervated by the mandibular nerve (CN V_3).

C. The palate is covered by mucosa containing glands and taste receptors.
1. The **mucosa of the hard palate** is innervated by the nasopalatine and the greater palatine nerves of the maxillary division of CN V.
2. The **mucosa of the soft palate** is innervated by the lesser palatine nerves of the maxillary division of V.
3. Postganglionic parasympathetic axons from cell bodies in the pterygopalatine ganglion innervate mucous glands in the palate. These axons reach the palate by hitchhiking on the palatine nerves of CN V_2.
4. **Taste fibers from mucosa of the soft palate** course with the lesser palatine nerves of CN V_2, pass through the pterygopalatine fossa, and follow the nerve of the pterygoid canal and the greater petrosal nerve of CN VII back to cell bodies in the geniculate ganglion. Central processes of cell bodies in the ganglion enter the CNS with the facial nerve.

X. The **temporal bone** contains 3 parts of the ear (Figure 8–18).
A. The **external ear and middle ear** are air-filled spaces; the **inner ear** is filled by connective tissue and fluid and consists of sacs and channels that contain hair cells in a variety of receptors.
1. External ear
a. The **external ear** is lined with skin and consists of the auricle, or pinna, which lies at the outer end of a short cartilaginous and bony tube, the external auditory meatus.
b. The **tympanic membrane** closes the medial aspect of the external auditory meatus. The depth of the external auditory meatus helps maintain a

Table 8–6. Muscles of the palate.

Action	Muscles Involved	Innervation
Tenses soft palate Opens auditory tube	Tensor veli palatini (palati)	Mandibular nerve of V_3
Elevates soft palate Opens auditory tube	Levator veli palatini	Vagus nerve (pharyngeal plexus)
Shortens soft palate	Musculus uvulae	Vagus nerve (pharyngeal plexus)
Narrows isthmus of fauces Elevates larynx and pharynx	Palatopharyngeus	Vagus nerve (pharyngeal plexus)

constant environment to permit the membrane to vibrate consistently in response to airborne stimuli.

  c. The external ear is innervated by auricular branches of CNs V, VII, IX, and X and by a cutaneous branch of the cervical plexus.

   (1) The anterior part of the external auditory meatus and most of the external surface of the tympanic membrane are innervated by the auriculotemporal nerve (CN V$_3$).

   (2) The auricular branches of the vagus and glossopharyngeal nerves innervate the posterior part of the external auditory meatus, part of the external surface of the tympanic membrane, and the posterior part of the pinna.

   (3) The posterior auricular nerve of the facial nerve innervates a small part of the external surface of the tympanic membrane and skin over the mastoid process.

   (4) The great auricular nerve from the C2 and C3 ventral rami of the cervical plexus contributes to the innervation of the inferior part of the pinna.

 2. Middle ear

  a. The **middle ear** is an air-filled chamber lined by mucosa. It is situated medial to the tympanic membrane (Figure 8–19).

  b. The middle ear contains the auditory tube, which connects the middle ear and the nasopharynx and provides a communication between the middle ear and the external environment.

  c. The middle ear **contains 3 ossicles**, which articulate at synovial joints.

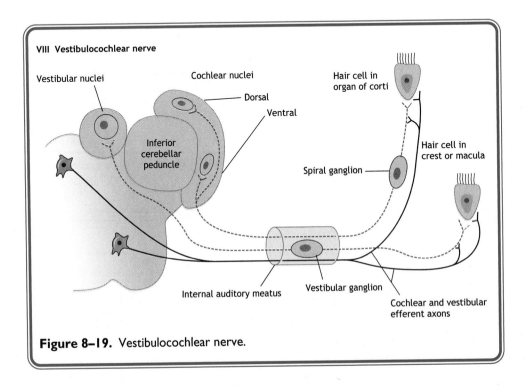

Figure 8–19. Vestibulocochlear nerve.

Table 8–7. Autonomics of head and neck.

Parasympathetics Preganglionic Axons Exit CNS In	Preganglionic Axons Exit Through	Preganglionic Axons Follow	Preganglionic Axons Synapse In	Postganglionic Axons Course With (Usually Branches of Trigeminal Nerve)	Structures Innervated
Parasympathetics					
Oculomotor nerve (III)	Superior orbital fissure	Oculomotor nerve (III) Inferior division	Ciliary ganglion (in orbit lateral to optic nerve)	Short ciliary nerves (V_1)	Sphincter pupillae muscle Ciliary muscle
Facial nerve (VII)	Internal auditory meatus	VII to chorda tympani of VII which joins lingual of V_3	Submandibular ganglion (hangs off of lingual n.)	Direct ganglion branches Rejoin lingual (V_3)	Submandibular gland Sublingual gland
Facial nerve (VII)	Internal auditory meatus	Greater petrosal n. in middle cranial fossa to nerve of pterygoid canal[a]	Pterygopalatine ganglion (in pterygopalatine fossa)	Pterygopalatine nerves to maxillary (V_2) to zygomatic to lacrimal (V_1) Nasal and palatine nerves of maxillary (V_2)	Lacrimal gland Mucous glands of nasal cavity, palate, pharynx
Glossopharyngeal nerve (IX)	Jugular foramen	Tympanic nerve thru middle ear to lesser petrosal nerve (exits through foramen ovale)	Otic ganglion (lies medial to V_3 below foramen ovale)	Auriculotemporal (V_3)	Parotid gland
Vagus nerve (X)	Jugular foramen	All branches of vagus distal to recurrent laryngeal nerve	Terminal ganglia (in wall of esophagus)	Nerves not named; lie in or near wall of structure innervated	Glands and smooth muscle in esophagus
Sympathetics					
Ventral root and ramus of TI	Intervertebral foramina	White rami to para-vertebral sympathetic trunk then up to	Superior cervical ganglion	Periarterial plexuses on carotid branches	Dilator pupillae muscle Superior tarsal muscle Sweat glands

[a]Formed by greater petrosal n. joining deep petrosal n. (postganglionic sympathetics)

221

(1) The **malleus** is attached to the inner surface of the tympanic membrane; the malleus articulates with the incus, and the incus articulates with the stapes.

(2) The **footplate of the stapes** is situated in the oval window against the fluid environment of the inner ear.

d. The **ossicles** function as amplifiers to overcome the **impedance mismatch** at the air-fluid interface of the middle ear and the inner ear.

e. Two skeletal muscles—the tensor tympani and stapedius—act to reduce the movement of the tympanic membrane and the ossicles.

(1) The tensor tympani attaches to the manubrium of the malleus and acts to reduce movement of the tympanic membrane. A branch of the mandibular nerve of CN V$_3$ innervates the tensor tympani.

(2) The stapedius attaches to the stapes near the incudostapedial joint and functions to partially disarticulate that joint to reduce movement of the stapes. The stapedius is innervated by the facial nerve.

HYPERACUSIS

CLINICAL CORRELATION

The **stapedius** contracts reflexively to protect the inner ear from high intensity vibrations. *A **lesion of the facial nerve**, if it includes the nerve to the stapedius, results in **hyperacusis**, an increased sensitivity to loud sounds.*

f. The mucosa of the middle ear, including mucosa covering the medial aspect of the tympanic membrane and mucosa of the auditory tube, is innervated by the tympanic nerve, a branch of the glossopharyngeal nerve.

3. **Inner ear**

a. The inner ear consists of a **labyrinth of interconnected sacs and channels** that contain endolymph, which bathes the hairs of the hair cells (see Figure 8–18).

b. **Endolymph** is unique because it has an inorganic ionic composition similar to intracellular fluid, but it is in an extracellular location. The ionic composition of endolymph is important for the function of hair cells.

c. The **utricle and saccule** contain hair cells in maculae that respond to linear acceleration and detect positional changes in the head relative to gravity.

d. The semicircular ducts contain hair cells in ampullary crests that respond to angular acceleration resulting from movements of the head.

e. The **cochlear duct** contains hair cells in the organ of Corti. These hair cells are situated on an elongated, flexible, basilar membrane.

(1) **High-frequency sounds** cause **maximum displacement of the basilar membrane and stimulation of hair cells** at the base of the cochlea.

(2) **Low-frequency sounds** cause **maximum displacement of the basilar membrane and stimulation of hair cells** at the apex of the cochlea.

B. The **vestibular part of the vestibulocochlear nerve** (CN VIII) innervates hair cells in the ampullary crests in the 3 semicircular ducts and hair cells in the utricular and saccular maculae (Figure 8–19).

C. The **cochlear part of the vestibulocochlear nerve** (CN VIII) innervates hair cells in the organ of Corti.

1. The vestibulocochlear nerve also contains axons of efferent neurons, which have their neuron cell bodies inside the CNS.
2. These efferent neurons function to regulate the sensitivity of the hair cells or the vestibular or cochlear nerve fibers that innervate them.

TYPES OF HEARING LOSS

Hearing losses may be conductive or sensorineural.
- *__Conductive hearing losses__ result from an interference of sound transmission through the external or middle ear. __Middle ear infections__ in children and __otosclerosis__ in adults are common causes of a conductive hearing loss.*
- *__Sensorineural hearing losses__ may result from a loss of hair cells in the cochlea or a lesion to the cochlear part of CN VIII or to CNS auditory pathways.*
- *__Presbycusis__ is a common cause of high-frequency sensorineural hearing loss in the elderly. It results from a progressive loss of hair cells at the base of the cochlea.*
- *__Ménière's disease__ is caused by an accumulation of endolymph (endolymphatic hydrops), which results in attacks of tinnitus (ringing) and vertigo.*

TESTS TO DETERMINE HEARING LOSS TYPE

- *The __Weber test__ may be used to determine whether a patient has a hearing loss but does not determine whether it is conductive or sensorineural.*
  – *In the Weber test, a vibrating tuning fork is placed in the midline of the skull or on the bridge of the nose; normally, vibrations are perceived equally in both ears.*
  – *Patients with a conductive hearing loss will hear the vibrations better on the side of the defective middle ear because vibrations reach the normal ear by both bone and air conduction, but they interfere with each other, making the normal ear less sensitive.*
  – *Patients with a sensorineural hearing loss will hear the vibrations better in the normal ear.*
- *The __Rinne test__ may be used to determine whether a patient has a conductive hearing loss.*
  – *In the Rinne test, the base of a vibrating tuning fork is placed on the mastoid process.*
  – *When the patient no longer hears the vibrations, the tuning fork is then placed next to the external ear. Normally, airborne vibrations are heard better than those conducted through bones of the skull (AC > BC) because of the efficiency of the middle ear, so the patient should again hear the vibrations.*
  – *If the patient has a conductive hearing loss, the tuning fork cannot be heard when placed next to the external ear (BC > AC).*

   D. The **facial nerve** (CN VII) contains preganglionic parasympathetic axons, skeletal motor axons, taste fibers, and general sensory fibers (Figure 8–11).
   1. The facial nerve enters the temporal bone through the internal auditory meatus, courses through the facial canal, and gives rise to branches that exit the temporal bone through the hiatus of the facial canal, the petrotympanic fissure, and the stylomastoid foramen.
   2. In the facial canal, the facial nerve bends sharply; at this genu is the **geniculate ganglion**, which contains the **cell bodies of taste fibers** from the anterior two-thirds of the tongue and palate and the cell bodies of general sensory neurons.
   3. The **2 main branches of the facial nerve** in the facial canal are the greater petrosal nerve and the chorda tympani. Both are mixed nerves that carry preganglionic parasympathetic axons and taste fibers (see Figure 8–11).
      a. The **greater petrosal nerve** branches at the geniculate ganglion and re-enters the middle cranial fossa through the hiatus of the facial canal.

(1) The greater petrosal nerve crosses the width of the upper part of the foramen lacerum and enters the pterygoid canal.

       (2) In the foramen lacerum, the greater petrosal is joined by the deep petrosal nerve and forms the nerve of the pterygoid canal.

       (3) Preganglionic parasympathetic axons in the nerve of the pterygoid canal synapse in the pterygopalatine ganglion (see Figure 8–11).

       (4) The taste fibers carried in the greater petrosal nerve and the nerve of the pterygoid canal innervate taste receptors in the soft palate.

    b. The **chorda tympani** branches from the facial nerve distal to the geniculate ganglion.

       (1) The chorda tympani leaves the facial canal, enters and passes through the middle ear between the malleus and the incus, and then exits the temporal bone through the petrotympanic fissure and enters the infratemporal fossa.

       (2) In the infratemporal fossa, the chorda tympani joins with the lingual nerve (CN V$_3$). The parasympathetic axons in the chorda tympani leave the lingual nerve and synapse in the submandibular ganglion (see Figure 8–11). Postganglionic parasympathetic axons from the ganglion innervate the submandibular and sublingual glands.

       (3) The chorda tympani also carries taste fibers from the anterior two-thirds of the tongue.

  4. The **nerve to the stapedius** arises between the greater petrosal nerve and the chorda tympani, enters the middle ear, and supplies the stapedius.

  5. The rest of the fibers of the facial nerve distal to the branch point of the chorda tympani and leave the temporal bone through the stylomastoid foramen.

    a. Most of the **fibers of the facial nerve** that exit the stylomastoid foramen innervate muscles of facial expression.

    b. These **skeletal motor fibers** enter the parotid gland and divide into temporal, zygomatic, buccal, mandibular, and cervical branches.

    c. The **posterior auricular nerve** innervates skin covering the mastoid process posterior to the external ear and a small part of the external ear and tympanic membrane. The posterior auricular nerve also innervates the occipital belly of the occipitofrontalis and small muscles that move the pinna.

FACIAL NERVE AND THE BLINK REFLEX

Fibers of the facial nerve form the motor limb of the corneal blink reflex.
- *The **blink reflex** uses sensory fibers of the ophthalmic division of CN V and skeletal motor fibers of the facial nerve (CN VII).*
- *Stimulation of the sensory fibers of CN V causes a direct and a consensual blink that results from bilateral contraction of the orbicularis oculi muscles.*

*A **lesion** to either the sensory fibers of the ophthalmic division of CN V or to the skeletal motor fibers of the facial nerve may disrupt the **blink reflex**.*

FACIAL NERVE LESIONS

Lesions of the facial nerve commonly occur in the facial canal and result in ***Bell's palsy***.
- *In patients with Bell's palsy, there is a weakness of muscles of facial expression on the side of the injured nerve.*
- *Patients experience weakness in the ability to shut the eye, flare a nostril, and wrinkle the forehead. There may also be a drooping of the corner of the mouth.*

• In addition, patients may have pain behind the external auditory meatus resulting from involvement of the general sensory fibers of the posterior auricular nerve.

If the **lesion of CN VII** is at the level of the geniculate ganglion, additional sensory and parasympathetic signs may be evident.

• There may be an alteration of taste sensations from the anterior two-thirds of the tongue and palate, a reduction in salivary secretions from the submandibular and sublingual glands, and a dry eye from a reduction in lacrimal secretions. **Hyperacusis** (a hypersensitivity to loud sounds) may result if the nerve to the stapedius is affected.

• As patients recover from a facial nerve lesion, they may experience **synkinesis**, which results from mis-directed regenerating motor axons.

  –**Crocodile tears**, or lacrimation while eating, results from parasympathetic axons to the sub-mandibular ganglion that aberrantly reinnervate the pterygopalatine ganglion.

Distal to the stylomastoid foramen, a **tumor of the parotid gland** may compress the muscular branches of the facial nerve as they traverse the gland and may result in a weakness of muscles of facial expression but no sensory deficits, hyperacusis, or alteration of glandular secretions.

Facial nerve lesions affecting muscles of facial expression also result in a loss of the motor limb of the blink reflex.

CLINICAL PROBLEMS

A construction worker suffers head trauma from a fall on the job. In the emergency room, clear, sticky fluid emerges through his nostrils. A week later, the patient notices that he has lost his appreciation of the smell of food and can no longer smell his morning coffee.

1. What bone may have been fractured to account for these symptoms?

  A. Temporal

  B. Maxillary

  C. Ethmoid

  D. Frontal

  E. Sphenoid

A 26-year-old woman has been back and forth to her physician complaining of frontal headaches that are resistant to medication. The patient has irregular menstrual cycles and notes that she seems to be losing her peripheral vision bilaterally. Magnetic resonance imaging reveals a tumor in the cranial cavity.

2. What structure might the tumor be compressing?

  A. An optic tract

  B. An optic nerve

  C. The lateral aspect of the optic chiasm

  D. The crossing fibers of the optic chiasm

  E. Structures passing through the superior orbital fissure

A 42-year-old woman is brought to the local health care center by her husband. The woman woke up with tingling and numbness in her lower face on the right, with brief episodes of sharp pain in the face when she attempted to chew food at breakfast. Later that morning, the pain recurred when she attempted to laugh at a cartoon in the newspaper.

3. What do the patient's symptoms suggest?

    A. Bell's palsy

    B. Trigeminal neuralgia

    C. Jugular foramen syndrome

    D. Horner's syndrome

    E. Berry aneurysm

Your patient complains of numbness in the mucosa covering the hard palate.

4. The sensory fibers conveying general sensation from the palate have their neuronal cell bodies in which ganglion?

    A. Trigeminal ganglion

    B. Geniculate ganglion

    C. Pterygopalatine ganglion

    D. Superior cervical ganglion

    E. Otic ganglion

Your patient comes to your office with the complaint that his face looks funny. The patient cannot wrinkle the forehead on one side, and the corner of the mouth droops. Your examination reveals an absence of a corneal blink reflex and a dry eye on the side of the weak muscles.

5. All of the patient's symptoms might be the result of:

    A. A tumor in the parotid gland

    B. A lesion of the facial nerve in the facial canal distal to the branch point of the chorda tympani

    C. A lesion of the pterygopalatine ganglion

    D. A thrombosis of the cavernous sinus

    E. A lesion at the level of the geniculate ganglion

6. The geniculate ganglion contains:

    A. Postganglionic parasympathetic neurons that innervate the lacrimal gland

    B. Postganglionic parasympathetic neurons, which innervate the sublingual gland

    C. Sensory cell bodies of neurons, which innervate skin covering the mastoid process

    D. Sensory cell bodies, which synapse with taste fibers from the anterior two-thirds of the tongue

    E. Sensory cell bodies of neurons, which carry general sensation for the anterior two-thirds of the tongue

A patient complains of nasal regurgitation of liquids as a result of a weakness in the ability to tense the soft palate during swallowing.

7. What else might you observe in the patient?

    A. Weakness in the ability to elevate the pharynx during swallowing

    B. Altered taste sensations from the oropharyngeal part of the tongue

    C. Saliva dripping from the corner of the mouth

    D. Deviation of the tongue toward the side of the weak muscle during protrusion

    E. Decreased ability to detect the temperature of a bolus of food on the anterior two-thirds of the tongue

A tumor compresses structures that traverse the jugular foramen.

8. What autonomic deficit might be observed in the patient?

    A. Decreased secretions of the parotid gland

    B. Horner's syndrome

    C. Inability to constrict the pupil in response to light

    D. Decreased secretions of the lacrimal gland

    E. Ptosis

9. What skeletal motor deficit might be observed in the same patient in Question 8?

    A. Weakness in the ability to elevate the hyoid bone during swallowing

    B. Weakness in the ability to protrude the tongue

    C. Dysphagia

    D. Weakness in the ability to turn the head toward the side of the tumor

    E. Hyperacusis

10. What sensory deficit might be observed in the same patient?

    A. Decreased sensation in skin covering the mandible

    B. Inability to feel the presence of a bolus of food on the posterior one-third of the tongue

    C. Alteration of taste in the anterior two-thirds of the tongue

    D. Inability to sense a corneal stimulus

    E. Decreased sensation in the nasal cavity

A 25-year-old man is brought to the emergency room because of an injury suffered in a street fight. He was struck in the face and now cannot close his mouth. He has intense pain in the right side of his jaw. The ramus of the mandible is displaced medially and superiorly, and the body is displaced inferiorly. A diagnosis of a fractured mandible is made.

11. Contractions of what muscle are responsible for the medial displacement of the ramus of the mandible?

    A. Temporalis, posterior fibers

    B. Masseter

C. Medial pterygoid

    D. Lateral pterygoid

    E. Buccinator

In the patient in Question 11, the nerve fibers that enter the mandible to innervate the mandibular teeth have been lesioned by the fracture.

12. What else might be noted by the patient as a result of the lesion?

    A. Inability to wrinkle the skin of the neck

    B. Weakness in the ability to elevate the hyoid bone

    C. Reduction in salivary secretions of the submandibular gland

    D. Alteration in taste from the anterior two-thirds of the tongue

    E. Numbness in skin of the chin

A teenager had a habit of picking blemishes on his face. One of the blemishes above the upper lip on the left became infected and spread from the face to the cavernous sinus where the infection formed a thrombolytic lesion of nerves that traverse the sinus.

13. Which of the following will not be affected by this lesion?

    A. Sensation in skin of the forehead

    B. The pupillary light reflex

    C. The ability to elevate the upper eyelid

    D. Lacrimal gland secretions

    E. Ability to abduct the eye

Your patient develops double vision that results from an inability to depress the eye when the eye is adducted.

14. What muscle might be weakened?

    A. Inferior rectus

    B. Inferior oblique

    C. Superior oblique

    D. Superior rectus

    E. Medial rectus

ANSWERS

1. The answer is C. A fracture of the cribriform plate of the ethmoid may tear central processes of olfactory neurons and the meninges of the olfactory bulb, resulting in anosmia and CSF rhinorrhea.

2. The answer is D. The patient appears to have a bitemporal hemianopsia caused by the tumor compressing the crossing fibers of the optic chiasm.

3. The answer is B. In patients with trigeminal neuralgia, pain radiates over the distribution of the maxillary and mandibular divisions of the trigeminal nerve. The pain associated with the neuralgia is frequently triggered by moving the mandible, smiling or yawning, or by cutaneous or mucosal stimulation.

4. The answer is A. Choices C, D, and E are autonomic ganglia; choice B mainly contains taste neurons of CN VII.

5. The answer is E. A lesion in Choices A, B, or C would produce some but not all of the symptoms. The cavernous sinus does not contain CN VII.

6. The answer is C. The ganglion contains sensory cell bodies of neurons that innervate skin covering the mastoid process (posterior auricular nerve). There are no parasympathetic neurons or synapses in this ganglion. Choice E cell bodies would be in the trigeminal ganglion.

7. The answer is E. The patient has a lesion of CN V_3.

8. The answer is A. Preganglionic parasympathetic axons of CN IX traverse the jugular foramen.

9. The answer is C. Motor fibers in CN X to the pharynx would be affected.

10. The answer is B. An inability to feel the presence of a bolus of food on the posterior one-third of the tongue results from compression of sensory fibers of CN IX.

11. The answer is C. The medial pterygoid elevates, protrudes, and causes medial deviation of the ramus of the mandible.

12. The answer is E. Numbness in the skin of the chin results from an injury to the mental branch of the inferior alveolar nerve.

13. The answer is D. The postganglionic parasympathetic axons that innervate the lacrimal gland arise from the pterygopalatine ganglion and enter the orbit without traversing the cavernous sinus.

14. The answer is C. The superior oblique is best tested by asking the patient to depress the adducted eye.

CHAPTER 9
NECK

I. The **visceral compartment** of the neck is anterior and extends from the base of the skull to the thoracic outlet (Figure 9–1).

   A. The **hyoid bone, the suprahyoid and infrahyoid muscles, pharynx, esophagus, larynx, and trachea** are in the visceral compartment of the neck.

   B. The **pretracheal layer of deep cervical fascia** encloses structures in the visceral compartment of the neck. Buccopharyngeal fascia is a continuation of pretracheal fascia, which surrounds the posterior aspect of the pharynx and esophagus.

II. The **vertebral compartment** of the neck is posterior and extends from the foramen magnum to the thoracic outlet (see Figure 9–1).

   A. The **cervical vertebrae, skeletal muscles** that attach to the cervical vertebrae, the **ventral rami** of the cervical plexus and brachial plexus, and the **vertebral arteries and veins** are in the **vertebral compartment of the neck**.

   B. The **prevertebral layer of deep cervical fascia** encloses structures in the vertebral compartment of the neck.

   C. The prevertebral layer is separated from the buccopharyngeal fascia by the retropharyngeal space.
      1. The retropharyngeal space is a potential space.
      2. Structures in the visceral compartment of the neck covered by pretracheal fascia glide against the prevertebral fascia during swallowing.
      3. An **infection in the retropharyngeal space** may spread inferiorly into the superior mediastinum.

   D. Both compartments of the neck are partially covered by 2 superficial muscles: the trapezius and the sternocleidomastoid (see Figure 9–1).
      1. An **investing layer of deep cervical fascia** encloses both compartments and splits to enclose the trapezius and the sternocleidomastoid.
      2. The **superficial cervical fascia** contains the platysma, a muscle of facial expression.

III. The **carotid sheath** contains neurovascular structures and is situated between the vertebral and visceral compartments of the neck (see Figure 9–1).

   A. The sheath is formed by **contributions from the pretracheal, prevertebral, and investing layers of deep cervical fascia.**

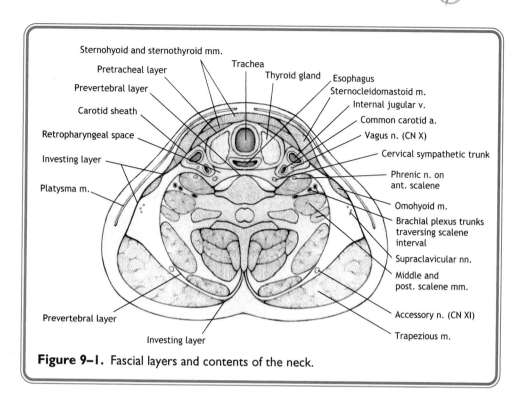

Figure 9–1. Fascial layers and contents of the neck.

B. The sheath contains the common carotid artery, which branches into the **internal and external carotid arteries** at the upper border of the thyroid cartilage.
  1. The **internal carotid artery** has no branches in the neck.
     a. The internal carotid artery enters the skull through the carotid canal; gives rise to the ophthalmic artery, which supplies the orbit, the retina, and part of the nasal cavity and face; and ends by branching into an anterior and a middle cerebral artery.
     b. The internal carotid artery conveys a periarterial plexus of postganglionic sympathetic axons from the superior cervical ganglion into the skull.
  2. The **external carotid artery** has 6 branches in the neck and ends posterior to the mandible by dividing into a superficial temporal artery and a maxillary artery (Figure 9–2).
     a. The **superior thyroid artery** supplies the thyroid gland and gives rise to the superior laryngeal artery, which passes through the thyrohyoid membrane to supply the laryngopharynx and larynx.
     b. The **ascending pharyngeal artery** arises from the posterior part of the external carotid and supplies the pharynx.
     c. The **lingual artery** passes deep to the mylohyoid to supply the tongue.
     d. The **facial artery** passes deep to the submandibular gland, crosses the body of the mandible, and supplies facial muscles and skin up to the medial corner of the eye.

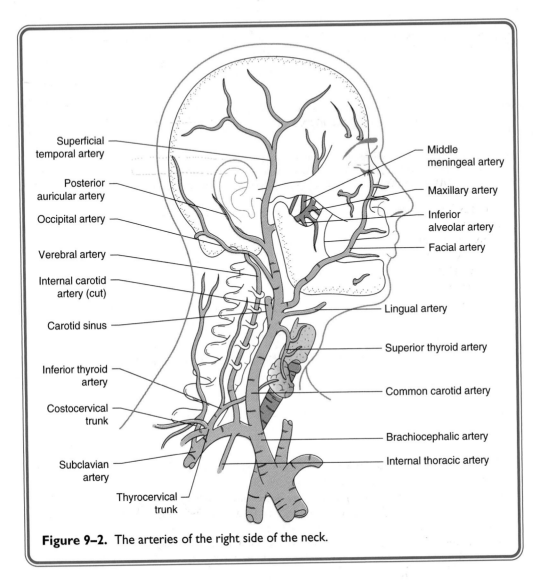

Figure 9–2. The arteries of the right side of the neck.

Labels on figure:
Superficial temporal artery
Posterior auricular artery
Occipital artery
Verebral artery
Internal carotid artery (cut)
Carotid sinus
Inferior thyroid artery
Costocervical trunk
Subclavian artery
Thyrocervical trunk
Middle meningeal artery
Maxillary artery
Inferior alveolar artery
Facial artery
Lingual artery
Superior thyroid artery
Common carotid artery
Brachiocephalic artery
Internal thoracic artery

   e. The **occipital artery** is crossed by the hypoglossal nerve and supplies the posterior neck and posterior scalp.
   f. The **posterior auricular artery** supplies the posterior scalp.
C. The **carotid body** is a chemoreceptor situated at the bifurcation of the common carotid artery that monitors arterial levels of oxygen and carbon dioxide.
D. The **carotid sinus** is a baroreceptor in the proximal part of the internal carotid artery that monitors arterial blood pressure.
   1. The carotid body and carotid sinus are innervated by branches of the glossopharyngeal and vagus nerves.

2. These nerves convey signals from the **baroreceptors and chemoreceptors** into the central nervous system and influence the parasympathetic and sympathetic innervation of the heart, lungs, and blood vessels.

BARORECEPTOR AND CHEMORECEPTOR REFLEXES

- *The **baroreceptor reflex** maintains blood pressure in response to changes in posture. **Disruption** of the baroreceptor reflex results in **orthostatic hypotension**, a decrease in blood pressure when the patient assumes an upright position.*
- *The **chemoreceptor reflex** maintains blood gases by adjusting respiration, cardiac output, and peripheral blood pressure. A **decrease in oxygen tension** (P_{O_2}) and an **increase in carbon dioxide tension** (P_{CO_2}) result in an **increase in respiration, heart rate, and peripheral blood pressure.***

E. The **internal jugular vein** courses through the sheath lateral to the common carotid artery.
  1. The internal jugular vein is formed just inferior to the jugular foramen at the junction of the inferior petrosal sinus and the sigmoid sinus.
  2. The internal jugular vein receives the facial, lingual, pharyngeal, and middle thyroid veins.

F. The **vagus nerve** courses through the length of the neck in the carotid sheath posterior to the internal jugular vein and the common carotid artery.

G. The **glossopharyngeal, accessory, and hypoglossal nerves** pass through the superior part of the sheath in their course to the pharynx, posterior triangle, and tongue, respectively.

IV. The cervical part of the **sympathetic trunk** lies posterior and medial to the carotid sheath and contains superior, middle, and inferior cervical sympathetic ganglia (see Figure 1–4).

A. **Superior Cervical Ganglion**
  1. The **superior cervical ganglion** receives preganglionic sympathetic axons mainly from the T1 segment of the spinal cord.
  2. It gives rise to postganglionic sympathetic axons, which supply sweat glands and vascular smooth muscle in the face and scalp, and innervate the dilator pupillae and superior tarsal muscles in the orbit.
  3. Postganglionic sympathetic axons from the superior cervical ganglion form **periarterial plexuses**, which course with the internal and external carotid arteries and their branches.
  4. It gives rise to gray rami that course with branches of the C1 through C4 spinal nerves supplying the neck.

B. **Inferior Cervical Ganglion**
  1. The **inferior cervical ganglion** is frequently fused with the first thoracic ganglion to form the stellate ganglion.
  2. The inferior cervical ganglion gives rise to gray rami that course with branches of the C7 and C8 ventral rami supplying the upper limb.

EXCESSIVE VASOCONSTRICTION

*A **stellate ganglion block** may be performed in patients who exhibit **excessive vasoconstriction** or sweating in the upper limb.*

HORNER'S SYNDROME

*A lesion of the stellate ganglion, the cervical part of the sympathetic trunk or the superior cervical ganglion may cause **Horner's syndrome**. Patients with Horner's syndrome have anhydrosis, ptosis, and miosis.*

V. The **posterior triangle** of the neck is bounded by the trapezius, the posterior border of the sternocleidomastoid, and the clavicle (Figures 9–1 and 9–3).

A. The **floor of the posterior triangle** contains the anterior scalene, middle scalene, posterior scalene, levator scapulae, and splenius capitis muscles, which are covered by prevertebral fascia.

B. The **roof of the posterior triangle** is formed by the investing layer of deep cervical fascia.

C. The **external jugular vein** crosses the sternocleidomastoid obliquely, pierces the investing fascia, and drains into the subclavian vein (Figure 9–5). The external jugular vein is formed by the union of the posterior auricular and the retromandibular veins.

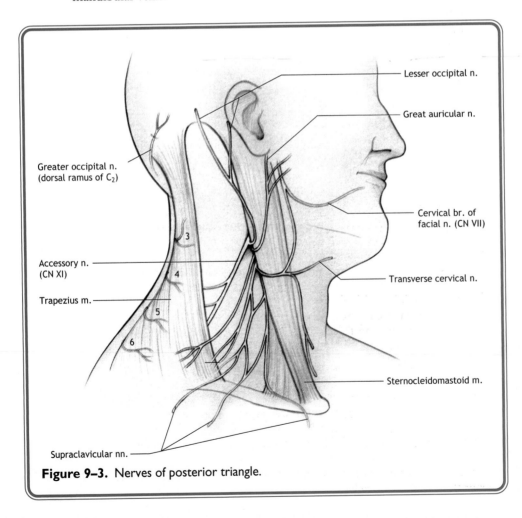

Figure 9–3. Nerves of posterior triangle.

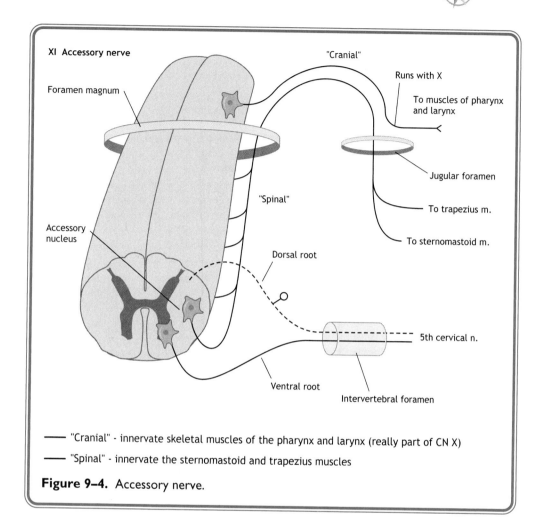

XI **Accessory nerve**

"Cranial"

Foramen magnum

Runs with X

To muscles of pharynx
and larynx

Jugular foramen

"Spinal"

To trapezius m.

Accessory
nucleus

Dorsal root

To sternomastoid m.

5th cervical n.

Ventral root

Intervertebral foramen

—— "Cranial" - innervate skeletal muscles of the pharynx and larynx (really part of CN X)

—— "Spinal" - innervate the sternomastoid and trapezius muscles

Figure 9–4. Accessory nerve.

D. The **subclavian vein** passes anterior to the phrenic nerve and the anterior scalene.
  1. The subclavian vein joins with the internal jugular vein to form a brachiocephalic vein posterior to the medial end of the clavicle.
  2. The right lymphatic duct and the thoracic duct drain into the right and left
  brachiocephalic veins, respectively, at their origins.
E. The **suprascapular, transverse cervical, and occipital arteries** course through
the posterior triangle.
  1. The suprascapular and transverse cervical arteries arise from the thyrocervical
  trunk of the subclavian, pass anterior to the anterior scalene and phrenic
  nerve, and cross the posterior triangle.
  2. They supply the trapezius, rhomboids, levator scapulae, and muscles that attach to the posterior aspect of the scapula.
  3. The occipital artery arises from the external carotid artery and passes through
  the apex of the posterior triangle.

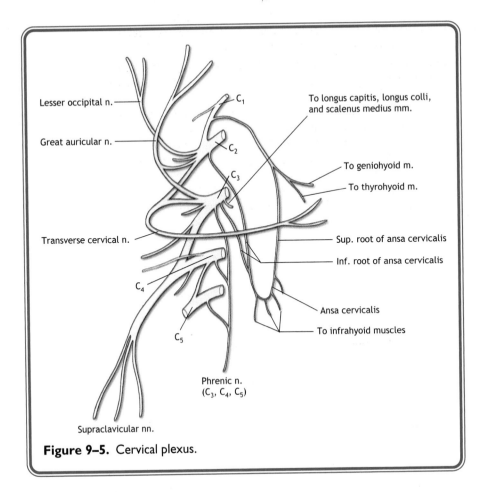

Lesser occipital n.

Great auricular n.

Transverse cervical n.

C_1

C_2

C_3

C_4

C_5

To longus capitis, longus colli, and scalenus medius mm.

To geniohyoid m.

To thyrohyoid m.

Sup. root of ansa cervicalis

Inf. root of ansa cervicalis

Ansa cervicalis

To infrahyoid muscles

Phrenic n. (C_3, C_4, C_5)

Supraclavicular nn.

Figure 9–5. Cervical plexus.

F. The **accessory nerve (CN XI)** (Figures 9–3 and 9–4) contains skeletal motor axons. Its motor axons arise from the spinal cord with the ventral roots of the first 4 or 5 cervical nerves, enter the subarachnoid space, and pass through the foramen magnum into the cranial cavity.
1. The accessory nerve exits the skull through the jugular foramen with the glossopharyngeal and vagus nerves.
2. It innervates the sternocleidomastoid and then crosses the posterior triangle and innervates the trapezius.

ACCESSORY NERVE LESIONS

*A **lesion of the accessory nerve** in the posterior triangle may result in weakness of the trapezius muscle. A patient may have difficulty elevating the scapula (shrugging the shoulder) and difficulty in laterally rotating the scapula during abduction of the arm.*
- *Lesions of the accessory nerve inferior to the jugular foramen may also result in weakness of the sternocleidomastoid. Patients may have a decreased ability to turn the chin to the side opposite the lesioned nerve.*

G. The **phrenic nerve** courses on the anterior surface of the anterior scalene muscle deep to prevertebral fascia (see Figures 9–1 and 9–5).

CLINICAL CORRELATION

H. The **ventral rami and the trunks of the brachial plexus and the subclavian artery** pass between the anterior and middle scalene muscles.

I. The **cutaneous branches of the cervical plexus** emerge posterior to the sternocleidomastoid approximately halfway between its sternal and mastoid attachments (see Figures 9–3 and 9–5).

1. These cutaneous branches arise from the ventral rami of the C1–4 spinal nerves and pierce the investing and superficial fascia.

2. They include the **great auricular nerve** (C2, C3), **lesser occipital nerve** (C2), **transverse cervical nerve** (C2, C3), and **supraclavicular nerves** (C2, C3, C4).

3. The cutaneous branches of the cervical plexus supply the following:
   a. Skin of the anterior and lateral neck, including skin over the angle of the mandible.
   b. Skin of the scalp posterior to the vertex of the skull, the coronal plane through the most superior aspect of the cranium.

VI. The **anterior triangle** of the neck is bounded by the anterior border of the sternocleidomastoid, the anterior midline, and the mandible.

A. The **anterior triangle** contains structures in the visceral compartment of the neck.

B. The **hyoid bone** is situated in the anterior triangle at the level of the C3 vertebra.

C. The **infrahyoid muscles** consist of 4 pairs of strap-like muscles that have attachments to the sternum, scapula, thyroid cartilage, and hyoid bone (Table 9–1).

1. The infrahyoid muscles include the sternohyoid, sternothyroid, superior and inferior bellies of omohyoid, and thyrohyoid.

2. The infrahyoid muscles are innervated by muscular branches of the cervical plexus by either the **ansa cervicalis** (sternohyoid, sternothyroid, superior and inferior bellies of omohyoid) or **C1 fibers** (thyrohyoid).
   a. The ansa cervicalis consists of a superior root from the C1 ventral ramus that hitchhikes with the hypoglossal nerve.

Table 9–1. Muscles that act on the hyoid bone.

Action	Muscles Involved	Innervation
Elevation	Mylohyoid	Nerve to mylohyoid (V_3)
	Geniohyoid	C1 fibers (run with CN XII)
	Stylohyoid	Cervical branch of facial (VII)
	Digastric	Anterior belly—n. to mylohyoid (V_3)
		Posterior belly—facial (VII)
Depression	Sternohyoid	C1, C2, and C3 (ansa cervicalis)
	Sternothyroid	C1, C2, and C3 (ansa cervicalis)
	Omohyoid	C1, C2, and C3 (ansa cervicalis)
	Thyrohyoid	C1 fibers (run with CN XII)

   b. The ansa cervicalis also has an inferior root from the C2 and C3 ventral rami that joins with the superior root anterior to the carotid sheath.

 D. The **suprahyoid muscles** have attachments to the styloid process, mandible, and hyoid bone.

  1. The suprahyoid muscles include the geniohyoid, mylohyoid, stylohyoid, and anterior and posterior bellies of the digastric.

  2. The suprahyoid muscles are innervated by branches of the trigeminal nerve (mylohyoid and anterior belly of digastric), facial nerve (CN VII) (stylohyoid and posterior belly of digastric), or C1 fibers from the cervical plexus (geniohyoid) (see Table 9–1).

VII. The **pharynx** is a fibromuscular tube that extends from the nasal cavity to the cricoid cartilage at the level of the sixth cervical vertebra. It consists of a nasopharynx, an oropharynx, and a laryngopharynx.

 A. The **nasopharynx** is superior to the soft palate and posterior to the nasal cavity.

  1. The lateral wall of the nasopharynx contains the opening of the auditory tube, levator veli palatini, tensor veli palatini, and salpingopharyngeus.

  2. The mucosa of the nasopharynx is innervated by the pharyngeal nerve, a branch of the maxillary division of V_2, and by branches of the glossopharyngeal nerve in the pharyngeal plexus.

 B. The **oropharynx** is situated between the soft palate and the tip of the epiglottis, posterior to the oral cavity.

  1. The lateral wall of the oropharynx contains the palatoglossal arches and the palatopharyngeal arches, which are separated by the tonsillar fossa.

   a. The palatoglossal arches contain the palatoglossal muscles; the palatopharyngeal arches contain the palatopharyngeus muscles.

   b. The tonsillar fossa contains the palatine tonsil and is crossed by the lingual branch of the glossopharyngeal nerve (CN IX).

  2. The floor of the oropharynx contains the posterior one-third of the tongue. Both taste and general sensations from the posterior one-third of the tongue are carried by the lingual branches of the glossopharyngeal nerve (CN IX) (Figure 9–6).

 C. The **laryngopharynx** extends from the epiglottis behind the larynx to the level of the cricoid cartilage. Inferior to the cricoid cartilage, the laryngopharynx is continuous with the esophagus.

  1. The lateral wall of the laryngopharynx contains the piriform recess. The internal branch of the superior laryngeal nerve of the vagus and the superior laryngeal artery course in the wall of the piriform recess.

  2. The mucosa of the laryngopharynx is innervated by the internal branch of the superior laryngeal nerve of the vagus (Figure 9–7).

 D. The **pharynx** consists of skeletal muscles that form an outer circular layer and an inner longitudinal layer (Table 9–2 and Figure 9–8).

  1. The circular layer is formed by 3 muscles—the superior, middle, and inferior constrictor muscles—that overlap and interdigitate in the posterior midline at the pharyngeal raphe.

   a. Contraction of the 3 constrictor muscles propels a bolus through the oropharynx and laryngopharynx during swallowing.

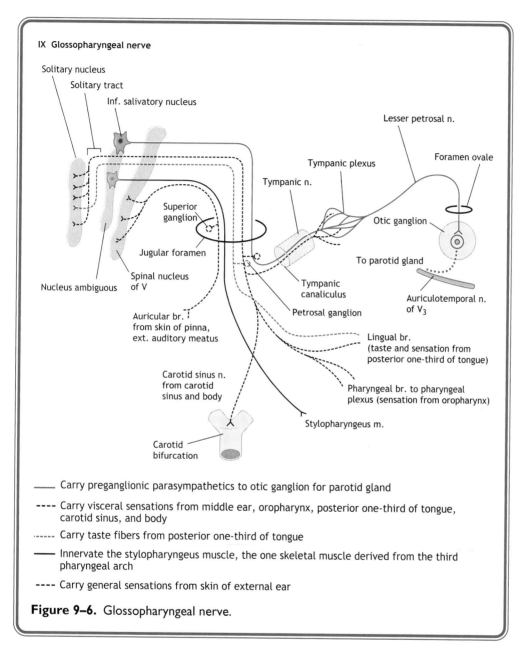

IX **Glossopharyngeal nerve**

Solitary nucleus

Solitary tract

Inf. salivatory nucleus

Lesser petrosal n.

Tympanic plexus

Tympanic n.

Foramen ovale

Superior ganglion

Otic ganglion

Jugular foramen

To parotid gland

Spinal nucleus of V

Nucleus ambiguous

Tympanic canaliculus

Petrosal ganglion

Auriculotemporal n. of V₃

Auricular br. from skin of pinna, ext. auditory meatus

Lingual br. (taste and sensation from posterior one-third of tongue)

Carotid sinus n. from carotid sinus and body

Pharyngeal br. to pharyngeal plexus (sensation from oropharynx)

Stylopharyngeus m.

Carotid bifurcation

_____ Carry preganglionic parasympathetics to otic ganglion for parotid gland

---- Carry visceral sensations from middle ear, oropharynx, posterior one-third of tongue, carotid sinus, and body

······ Carry taste fibers from posterior one-third of tongue

—— Innervate the stylopharyngeus muscle, the one skeletal muscle derived from the third pharyngeal arch

---- Carry general sensations from skin of external ear

Figure 9–6. Glossopharyngeal nerve.

    b. The inferior constrictor becomes continuous with the esophagus at the level of the cricoid cartilage.

  2. The inner longitudinal muscle layer is formed by 3 longitudinal muscles, the salpingopharyngeus, palatopharyngeus, and stylopharyngeus, that expand and insert into the pharyngeal wall. The 3 longitudinal muscles elevate the pharynx during swallowing.

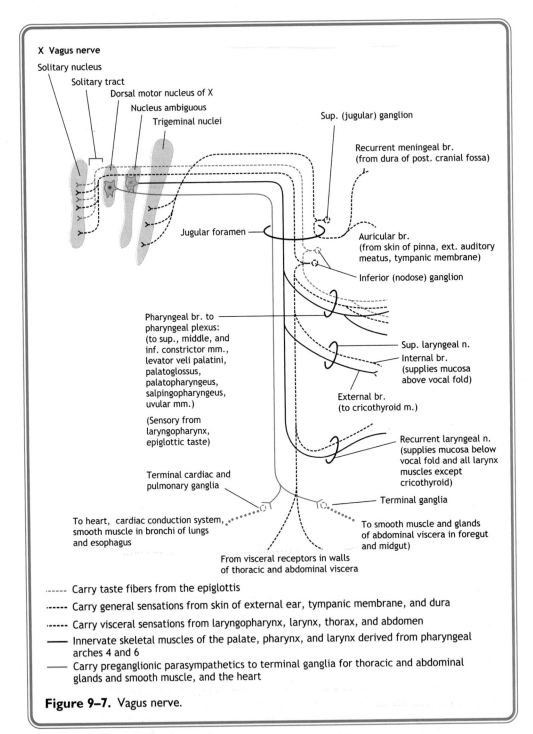

Figure 9–7. Vagus nerve.

Table 9–2. Muscles of the pharynx.

Action	Muscles Involved	Innervation
Longitudinal muscles		
Elevates pharynx and larynx	Stylopharyngeus	Glossopharyngeal nerve
Elevates pharynx	Salpingopharyngeus	Vagus nerve (pharyngeal plexus)
Elevates pharynx and larynx	Palatopharyngeus	Vagus nerve (pharyngeal plexus)
Circular muscles		
Constricts pharynx	Superior constrictor	Vagus nerve (pharyngeal plexus)
Constricts pharynx	Middle constrictor	Vagus nerve (pharyngeal plexus)
Constricts pharynx	Inferior constrictor	Vagus nerve (pharyngeal plexus), External br. of superior laryngeal nerve of X Recurrent laryngeal nerve

3. All of the **muscles of the pharynx** are innervated by branches of the vagus nerve through the pharyngeal plexus except for the stylopharyngeus, which is innervated by the glossopharyngeal nerve.
4. The inferior constrictor is also innervated by the external laryngeal nerve and the recurrent laryngeal branches of the vagus.

E. The **esophagus** begins inferior to the inferior constrictor at the level of the C6 vertebra.

F. The **pharynx** is supplied by branches of neurovascular structures that enter between the superior constrictor and the base of the skull, between the constrictors, or between the inferior constrictor and the esophagus (Figure 9–8).
   1. The auditory tube, the levator veli palatini (palati) muscle, and the ascending palatine artery enter the nasopharynx between the superior constrictor and the base of the skull.
   2. The stylopharyngeus, the stylohyoid ligament, and the glossopharyngeal nerve (CN IX) enter the oropharynx between the superior and middle constrictors.
   3. The internal branch of the superior laryngeal nerve of the vagus (CN X) and the superior laryngeal artery and vein enter the laryngopharynx between the middle and inferior constrictors.
   4. The inferior laryngeal nerve of CN X and the inferior laryngeal artery enter the pharynx between the inferior constrictor and the esophagus.

G. The **glossopharyngeal nerve** (CN IX) contains skeletal motor axons, preganglionic parasympathetic axons, taste fibers, and general sensory fibers (Figure 9–6).
   1. The glossopharyngeal nerve traverses the jugular foramen, and its skeletal motor axons innervate a single muscle: the stylopharyngeus.

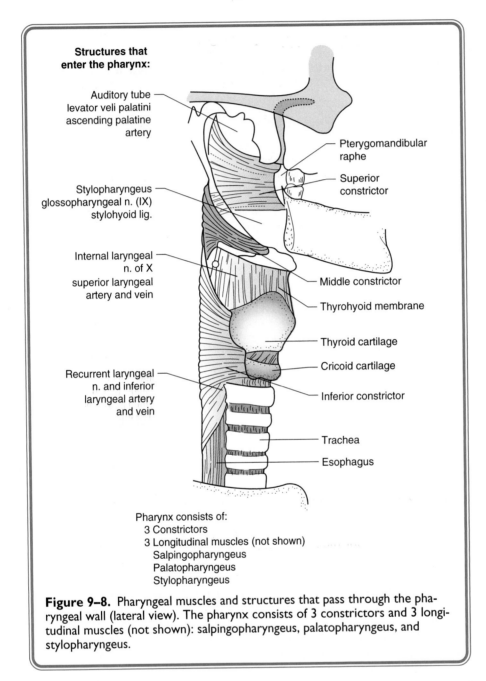

Structures that enter the pharynx:

- Auditory tube
- levator veli palatini
- ascending palatine artery
- Stylopharyngeus
- glossopharyngeal n. (IX)
- stylohyoid lig.
- Internal laryngeal n. of X
- superior laryngeal artery and vein
- Recurrent laryngeal n. and inferior laryngeal artery and vein

- Pterygomandibular raphe
- Superior constrictor
- Middle constrictor
- Thyrohyoid membrane
- Thyroid cartilage
- Cricoid cartilage
- Inferior constrictor
- Trachea
- Esophagus

Pharynx consists of:
3 Constrictors
3 Longitudinal muscles (not shown)
Salpingopharyngeus
Palatopharyngeus
Stylopharyngeus

Figure 9–8. Pharyngeal muscles and structures that pass through the pharyngeal wall (lateral view). The pharynx consists of 3 constrictors and 3 longitudinal muscles (not shown): salpingopharyngeus, palatopharyngeus, and stylopharyngeus.

2. The preganglionic parasympathetic axons in the glossopharyngeal nerve participate in the innervation of the parotid gland.
   a. The preganglionic parasympathetic axons course in the tympanic nerve and in the lesser petrosal nerve.
   b. The lesser petrosal nerve passes through the foramen ovale and synapses in the otic ganglion, located in the infratemporal fossa just below the foramen ovale.

    c. Postganglionic parasympathetic axons from the otic ganglion join with the auriculotemporal nerve to reach the parotid gland.

3. Through the pharyngeal plexus, which also contains vagus nerve fibers and postganglionic sympathetics, the glossopharyngeal nerve innervates the mucous membranes of the posterior wall of the oropharynx and the nasopharynx inferior to the entrance of the auditory tube.

    a. The lingual branch of CN IX conveys general sensation and taste from the posterior one-third of the tongue.

    b. A tonsillar branch supplies the mucosa of the palatoglossal and palatopharyngeal arches and the intervening tonsillar fossa.

4. The carotid sinus branch of CN IX innervates the carotid body and the carotid sinus.

H. The **gag reflex** uses the sensory fibers of the glossopharyngeal nerve (IX) in the oropharyngeal mucosa and motor fibers in the vagus nerve. Stimulation of CN IX results in a bilateral contraction of the pharyngeal musculature and an elevation of the soft palate.

JUGULAR FORAMEN SYNDROME

*Lesions of the glossopharyngeal nerve may occur in conjunction with the vagus nerve and the accessory nerve in **jugular foramen syndrome**. They are reliably diagnosed only when accompanied by a loss of the gag reflex.*

*The **lingual branch of CN IX** may be lesioned as it traverses the floor of the tonsillar fossa during **tonsillectomy** procedures. This lesion results in a loss of all sensation from the posterior one-third of the tongue.*

MIDDLE EAR INFECTIONS

Middle ear infections *may affect the preganglionic parasympathetic axons of CN IX, which synapse in the otic ganglion, and result in reduced parotid gland secretions. A reduction in parotid secretions into the oral cavity is difficult to evaluate because the submandibular and sublingual salivatory glands, which are innervated by the facial nerve, contribute to the volume of saliva.*

I. The **vagus nerve** (CN X) contains skeletal motor axons, preganglionic parasympathetic axons, taste fibers, and general sensory fibers (Figure 9–7).

1. The vagus nerve traverses the jugular foramen and innervates all of the muscles of the palate except for the tensor veli palatini, all of the muscles of the pharynx except for the stylopharyngeus, and all of the muscles of the larynx.

2. The vagus nerve innervates the mucosa at the root of the tongue and in the laryngopharynx and larynx.

3. The vagus nerve carries visceral sensations other than pain from thoracic and abdominal viscera.

4. The vagus nerve also conveys preganglionic parasympathetic axons to terminal ganglia in thoracic and abdominal viscera.

VAGUS NERVE LESIONS

*Complete **lesions of the vagus nerve** commonly result in a weakness of palate, pharyngeal, and laryngeal muscles.*

• *Weakness of the levator veli palatini may result in a drooping of the palate on the side of the injured nerve and a deviation of the uvula to the side opposite the lesioned nerve. Patients may also experience nasal speech and nasal regurgitation of liquids during swallowing.*

- *Weakness of the pharyngeal constrictors may result in **dysphagia**, or difficulty in swallowing.*
- *Lesions of the vagus nerve that include the laryngeal nerves may result in a weakness of all laryngeal muscles on the affected side. The vocal cord will assume a fixed position midway between abduction and adduction, resulting in speech that is hoarse and weak. Lesions of the pharyngeal branches and laryngeal nerves may also result in a loss of the motor limb of the gag reflex and the cough reflex, respectively.*

VIII. The **larynx** consists of cartilages and skeletal muscles that open and close the airway and allow phonation.

    A. The **vestibule of the larynx** begins at the inlet of the larynx behind the epiglottis and extends to the vestibular folds.

    B. The **ventricle is a narrow ellipse-shaped space** situated between the vestibular folds and the vocal folds.

    C. The **infraglottic space** is inferior to the vocal folds.
        1. The **rima glottis** is the opening between the vocal folds.
        2. The **glottis** consists of the rima glottidis and the vocal folds.

    D. The **larynx** contains 2 large single cartilages—the cricoid cartilage and the thyroid cartilage—and a pair of arytenoid cartilages. The cricoid, thyroid, and arytenoid cartilages articulate with one another at synovial joints (Figure 9–9a, b, c).
        1. The thyroid cartilage is *V*-shaped and articulates inferiorly with the ring-shaped cricoid cartilage. Inferiorly, the cricoid cartilage is continuous with the trachea.
        2. The arytenoid cartilages lie on the posterosuperior aspect of the cricoid cartilage and have muscular and vocal processes.
            a. The vocal ligaments extend anteriorly from the vocal processes of the arytenoids and attach to the posterior aspect of the thyroid cartilage.
            b. The vocal ligament is covered by a thyroarytenoid muscle, which is then covered by mucosa; these 3 structures form the vocal fold or vocal cord (Figure 9–9a, b).

    E. **Two pairs of skeletal muscles of the larynx** have antagonistic actions on the vocal ligaments (Table 9–3).
        1. The **lateral cricoarytenoid muscles** adduct the vocal ligaments; the **posterior cricoarytenoid muscles** abduct the vocal ligaments; both muscles rotate the arytenoid cartilages on the cricoid cartilage (see Figure 9–9a, b).
            a. Full adduction of the vocal ligaments causes the vocal folds to meet in the midline, closing the glottis during swallowing, defecation, and urination.
            b. When the vocal ligaments are partially adducted, air passing between the vocal folds causes the folds to vibrate during phonation.
        2. The **thyroarytenoid muscles** relax the vocal ligaments, and the **cricothyroid** muscles tense the vocal ligaments (Figure 9–9a, b, c).
            a. The thyroarytenoid muscles relax the ligaments by pulling the arytenoid cartilages closer to the thyroid cartilage.
            b. The vocalis muscle forms the medial part of the thyroarytenoid and adjusts the tension in small segments of the vocal ligament.
            c. The cricothyroid muscles tense the vocal ligaments by rocking the superior aspect of the thyroid anteriorly at its articulation with the cricoid, increasing the distance between these 2 cartilages.

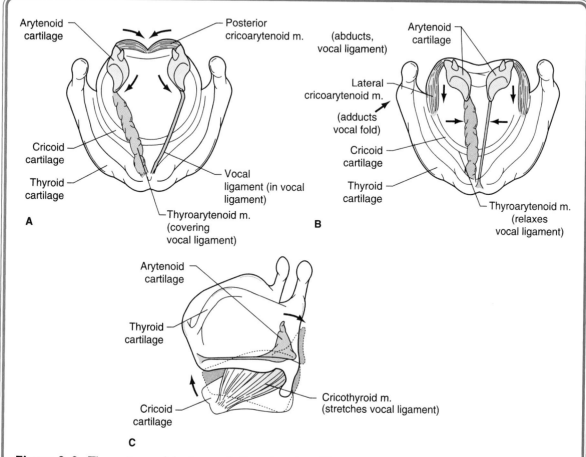

Figure 9–9. Three views of the larynx A: Superior view. The posterior cricoarytenoid muscles are the only muscles that abduct the vocal ligaments. B: Superior view. The muscles that adduct the vocal ligaments are the lateral cricoarytenoids. C: Lateral view. Contraction of the cricothyroid muscle elevates the arch of the cricoid cartilage and stretches the vocal ligaments.

F. All of the **muscles of the larynx** are innervated by the **inferior laryngeal nerve**, a branch of the recurrent laryngeal nerve of the vagus, except for the cricothyroid muscle, which is innervated by the external branch of the superior laryngeal nerve (Figure 9–7).
1. The **internal branch of the superior laryngeal nerve** innervates laryngeal mucosa above the vocal folds.
2. The **recurrent laryngeal nerve** innervates mucosa below the vocal folds.

SUPERIOR LARYNGEAL NERVE LESIONS

Lesions of the superior laryngeal nerve are largely asymptomatic because its fibers are mainly sensory. If the motor fibers to the cricothyroid are affected in a lesion of the external branch, a patient may experience mild hoarseness and a slight decrease in vocal strength, with a tendency to produce monotonous speech.

CLINICAL CORRELATION

Table 9–3. Muscles of the larynx.

Action	Muscles Involved	Innervation
Adducts vocal ligament Assists in closing glottis	Lateral cricoarytenoid	Recurrent laryngeal nerve of X
Abducts vocal ligament Opens glottis	Posterior cricoarytenoid	Recurrent laryngeal nerve of X
Relaxes vocal ligament	Thyroarytenoid	Recurrent laryngeal nerve of X
Regulates tension of segments of vocal ligament	Vocalis	
Tenses vocal folds	Cricothyroid	External br. of superior laryngeal nerve of X
Closes glottis by adducting arytenoid cartilages	Transverse Arytenoid	Recurrent laryngeal nerve of X
Narrows inlet of larynx	Aryepiglotticus	Recurrent laryngeal nerve of X

RECURRENT LARYNGEAL NERVE LESIONS

*Both **recurrent laryngeal nerves** are susceptible to injury in surgical procedures involving the thyroid gland. Lesions of a recurrent laryngeal nerve result in a fixed vocal cord and transient hoarseness.*
- *The left recurrent laryngeal nerve is injured more frequently than the right because of its course through the superior mediastinum.*
- *In the mediastinum, the left recurrent laryngeal nerve hooks around the arch of the aorta and might be compressed by an aortic aneurysm.*
- *The right recurrent laryngeal nerve is found only in the neck, where it hooks around the right subclavian artery.*

IX. The **trachea** begins inferior to the cricoid cartilage at the level of the C6 vertebra (see Figure 9–8).

   A. The trachea extends inferiorly into the mediastinum and ends by bifurcating into the left and right primary bronchi at the disk between the T4 and T5 vertebrae.

   B. The trachea consists of cartilage rings that are incomplete posteriorly and a posterior wall of smooth muscle.

   C. The **lobes and isthmus of the thyroid gland** partially cover the trachea.
   1. The lobes of the thyroid are lateral to the trachea; the isthmus, which passes anterior to the second or third tracheal rings, interconnects the lobes.
   2. The thyroid gland produces and secretes triiodothyronine (T_3), thyroxine (T_4), and calcitonin.

   D. A **pair of superior parathyroid glands and a pair of inferior parathyroid glands** are situated posterior to the lobes of the thyroid. The parathyroid glands produce and secrete parathyroid hormone.

E. The **superior and inferior thyroid arteries** supply the thyroid and parathyroid glands.

    1. The external branch of the superior laryngeal nerve of the vagus courses with the superior thyroid artery.

    2. The recurrent laryngeal nerve courses in the groove between the trachea and the esophagus in close proximity to the inferior thyroid artery.

CLINICAL PROBLEMS

Clinical features match. Match the clinical feature in Questions 1–25 with the appropriate nerve or neural structure. Use the following choices (note that some may be used once, more than once, or not at all):

  A. Trigeminal nerve

  B. Oculomotor nerve

  C. Glossopharyngeal nerve

  D. Vagus nerve

  E. Facial nerve

  F. Superior cervical ganglion

  G. Hypoglossal nerve

  H. Abducens nerve

  I. Trochlear nerve

  J. MORE THAN ONE OF THE ABOVE CHOICES IS CORRECT

1. Patient has a drooping upper eyelid.

2. The patient's uvula deviates away from the side of a lesioned nerve.

3. Patient can shut both eyes but cannot feel a corneal stimulus.

4. Patient has altered sensation from mucosa of the nasal cavity.

5. Patient cannot shut an eye.

6. Patient has an inability to adduct an eye.

7. Patient cannot keep saliva from dripping out of a corner of the mouth.

8. Patient is hypersensitive to loud sounds.

9. Patient has double vision.

10. Patient has difficulty swallowing, but feels a tongue depressor touching oropharynx.

11. Patient has pain in skin over the mastoid process.

12. The diameters of the pupils are not equal.

13. Patient cannot change the shape of the lens in response to a near stimulus.

14. Patient has no cough reflex.

15. Patient has hoarseness.

16. Patient's face is dry on one side.

17. Patient cannot feel tongue depressor touching anterior two-thirds of the tongue.

18. Patient has a reduction in salivary gland secretions.

19. Patient can stick the tip of the tongue inside a cheek, but cannot feel the mucosa.

20. Patient has a lesion of parasympathetic axons.

21. Patient has internal strabismus.

22. Patient has diplopia when attempting to depress an eye from the adducted position.

23. Patient has diplopia during attempted convergence.

24. Patient has reduced peristalsis in the pharynx.

25. Patient has altered taste sensations from the tongue.

A male teenager is admitted to the emergency room with a stab wound of the neck. The wound is several inches inferior to the mastoid process, just posterior to the sternocleido-mastoid. The next week, the patient complains of weakness in the ability to shrug his shoulder on the side of the injury.

26. What else might be observed in the patient?

    A. Weakness in the ability to protract the scapula

    B. Weakness in the ability to abduct his arm above his head to comb his hair

    C. Weakness in the ability to depress the hyoid bone

    D. Weakness in the ability to turn the face to the opposite side of the injury

    E. Weakness in the ability to depress the mandible

Your patient has suffered a stab wound to the neck that lacerates structures entering the pharynx through the thyrohyoid membrane.

27. Which of the following symptoms might the patient have?

    A. Weakness in the ability to adduct the vocal cord

    B. Decreased ability to detect a foreign body in contact with the mucosa below the vocal fold

    C. Weakness in the ability to tense the vocal ligament

    D. Hoarseness

    E. Decreased ability to detect a foreign body in contact with the mucosa in the piriform recess

A tonsillectomy procedure to remove the palatine tonsil lesions a nerve in the lateral wall of the tonsillar fossa.

28. What might an examination reveal?

    A. The patient may have an absence of the gag reflex on the side of the lesion.

    B. The patient's tongue may deviate on protrusion toward the side of the tonsil-lectomy.

C. The patient may have a deviated uvula.

   D. The patient may have hyperacusis.

   E. The patient may not be able to open the auditory tube when swallowing.

A thyroidectomy procedure lesions a nerve that enters the pharynx at the junction of the inferior constrictor and the esophagus.

29. Which of the following functions will be affected by the lesion?

   A. The ability to elevate the larynx during swallowing

   B. The ability to elevate the palate during swallowing

   C. The ability to abduct the vocal fold

   D. The ability to contract the middle constrictor during swallowing

   E. The ability to detect the presence of a foreign body in the mucosa of the vestibule of the larynx

A woman undergoes surgery for removal of a swelling in the left lobe of her thyroid gland. Postoperative examination reveals that her voice is hoarse.

30. What nerve may have been affected by the surgical procedure?

   A. Superior laryngeal nerve

   B. Internal laryngeal nerve

   C. Recurrent laryngeal nerve

   D. External laryngeal nerve

Your patient presents with hoarseness that results from an inability to abduct or adduct the right vocal cord. The patient is able to swallow normally, and the palate elevates symmetrically during testing of the gag reflex.

31. A complete lesion of which of the following neural structures might account for the patient's symptoms?

   A. Glossopharyngeal nerve

   B. Vagus nerve

   C. Superior laryngeal nerve

   D. Recurrent laryngeal nerve

   E. Internal laryngeal nerve

Your patient exhibits a decrease in parotid gland secretions on the right, and stimulation of the oropharyngeal mucosa on the right does not result in elevation of the palate or contraction of pharyngeal constrictors.

32. Which of the following might account for these symptoms?

   A. A tumor in the posterior cranial fossa compressing nerve fibers that traverse the jugular foramen

   B. A lesion of the glossopharyngeal nerve as its passes between the superior and middle pharyngeal constrictors

   C. A tonsillectomy procedure that lacerates the glossopharyngeal nerve in the tonsillar fossa

D. An infection in the middle ear

E. A thyroidectomy procedure that lesions the right recurrent laryngeal nerve

ANSWERS

1. The answer is J. MORE THAN ONE OF THE ABOVE CHOICES IS CORRECT. A lesion of either the superior cervical ganglion or the oculomotor nerve may cause a ptosis.

2. The answer is D. Vagus nerve. Weakness of a levator veli palatini causes the uvula to deviate away from the side of a lesioned vagus nerve.

3. The answer is A. Trigeminal nerve. Ciliary branches of the ophthalmic division convey sensation from the cornea.

4. The answer is A. Trigeminal nerve. The maxillary division and to a lesser degree the ophthalmic division supply the nasal cavity.

5. The answer is E. Facial nerve. CN VII innervates the orbicularis oculi muscle that shuts the eye.

6. The answer is B. Oculomotor nerve. CN III innervates all ocular muscles that adduct the eye.

7. The answer is E. Facial nerve. CN VII innervates the orbicularis oris muscle that closes the mouth.

8. The answer is E. Facial nerve. CN VII innervates the stapedius muscle that limits movement of the stapes.

9. The answer is J. MORE THAN ONE OF THE ABOVE CHOICES IS CORRECT. A lesion of CNs III, IV, or VI may cause double vision.

10. The answer is D. Vagus nerve. Weakness of the pharyngeal constrictors may cause dysphagia.

11. The answer is E. Facial nerve. The posterior auricular branch of CN VII innervates skin over the mastoid process.

12. The answer is J. MORE THAN ONE OF THE ABOVE CHOICES IS CORRECT. A lesion of either the superior cervical ganglion or the oculomotor nerve may cause either a constricted or a dilated pupil, respectively.

13. The answer is B. Oculomotor nerve. CN III innervates the ciliary muscle that allows the lens to round up for near vision.

14. The answer is D. Vagus nerve. Sensory and motor fibers of CN X participate in the cough reflex.

15. The answer is D. Vagus nerve. Weakness of the laryngeal muscles innervated by CN X (recurrent laryngeal nerve) results in hoarseness.

16. The answer is F. Superior cervical ganglion. A dry face is one of the signs of Horner's syndrome.

17. The answer is A. Trigeminal nerve. The lingual nerve of CN V3 conveys general sensation from the anterior two-thirds of the tongue.

18. The answer is J. MORE THAN ONE OF THE ABOVE CHOICES IS CORRECT. Both CN VII and CN IX innervate salivary glands.

19. The answer is A. Trigeminal nerve. The buccal nerve of CN V3 conveys general sensation from oral mucosa.

20. The answer is J. MORE THAN ONE OF THE ABOVE CHOICES IS CORRECT. CNs III, VII, IX, and X carry parasympathetic axons.

21. The answer is H. Abducens nerve. Loss of the lateral rectus causes a medial deviation of the eye.

22. The answer is I. Trochlear nerve. The superior oblique is the depressor of the eye from the adducted position. The inferior rectus (CN III) depresses the eye from the abducted position.

23. The answer is B. Oculomotor nerve. CN III innervates the medial rectus muscles that contract to converge the eyes during the near response.

24. The answer is D. Vagus nerve. Weakness of the pharyngeal muscles may cause dysphagia and decreased peristalsis.

25. The answer is J. MORE THAN ONE OF THE ABOVE CHOICES IS CORRECT. CNs VII and IX convey taste from the tongue.

26. The answer is B. The accessory nerve is most commonly lesioned in the posterior triangle and results in a weakness of the trapezius muscle, which acts to elevate and rotate the scapula during abduction of the arm.

27. The answer is E. The internal branch of the superior laryngeal nerve innervates mucosa of the laryngopharynx, and mucosa of the larynx above the vocal folds.

28. The answer is A. The patient may have an absence of the gag reflex caused by a lesion of CN IX.

29. The answer is C. The ability to abduct the vocal fold will be affected; the inferior laryngeal nerve innervates all of the muscles of the larynx except cricothyroid.

30. The answer is C. Lesions of a recurrent laryngeal nerve result in a fixed vocal cord and transient hoarseness.

31. The answer is D. The recurrent laryngeal nerve innervates all of the muscles that abduct or adduct the vocal ligaments.

32. The answer is A. The patient has a complete lesion of all of the fibers in the glossopharyngeal nerve.

INDEX

Notes

Notes

Notes

Notes

Notes

Notes

Notes

Notes